Intermolecular Forces

Intermolecular Forces

T. Kihara

Department of Physics
University of Tokyo

Translated by

S. Ichimaru

Department of Physics
University of Tokyo

JOHN WILEY & SONS
Chichester • New York • Brisbane • Toronto

Bunshikanryoku by Taro Kihara
Originally published in Japanese
by Iwanami Shoten, Publishers, Tokyo, 1976

Translated by Setsuo Ichimaru

Library of Congress Cataloging in Publication Data:

Kihara, Taro.
Intermolecular forces.

Includes index.
1. Molecular theory. I. Title.
QD461.K425 541′226 77-12353
ISBN 0 471 99583 5

Typeset in IBM Press Roman by
Preface Ltd, Salisbury, Wilts
Printed and bound in Great Britain
by Pitman Press Ltd., Bath

Contents

Preface to the Original Japanese Edition

The principal subjects of this book are intermolecular forces; problems involving intermolecular forces, such as quantum-mechanical perturbation theory, statistical mechanics and the kinetic theory of gases, and structures of molecular crystals, are treated. Assuming only elementary knowledge of quantum mechanics to the readers, we develop each subject matter from fundamental principles. We thus begin with expositions on the basics, such as formulas in perturbation theory, canonical and the grand canonical ensembles, the Boltzmann equation, and the space groups representing symmetries in crystal structures. We then clarify how those fundamental concepts coupled with the intermolecular forces may correspond to experimental results.

Different molecules have their own distinct characters, or individualities. Chemistry is concerned with a systematic study of such individualities. The spirit of this book throughout is respect of the individualities. Emphasizing different features, however, does not go along with clear description and presentation; a 'limitation' must be placed. Thus in the equation of state for gases, the molecular shapes will be taken into account only to an extent describable in terms of the shapes of convex bodies; in the structures of molecular crystals, models involving magnetic multipoles will be regarded as a limit within which individualities of intermolecular forces may be considered.

Considerable effort has been placed in this book to provide a most systematic presentation of intermolecular forces as an interdisciplinary subject between theoretical physics and chemistry, namely, the chemical physics of intermolecular forces. I hope that the book will be a suitable one for the original aims of the Iwanami Concise Textbook Series.

I wish to dedicate this book, with my special gratitudes, to Professor Masao Kotani and Professor Joseph O. Hirschfelder.

February 1976 Taro Kihara

Preface to the English Edition

In this book I do not intend to give a complete set of references. Only closely related papers are cited in footnotes and in the *Bibliography*. In the English edition, however, *Bibliographical Notes* is newly added on publisher's advice.

I should like to express here my sincere thanks to Professor S. Ichimaru for his excellent translation of the book.

November 1977 Taro Kihara

Chapter 1

Introduction

1.1 Attractive and repulsive forces between molecules

Gases such as air and argon are liquefied at sufficiently low temperatures: this is a manifestation of attractive forces between molecules. In a gaseous state, the motion of molecules is vigorous; they move almost freely, overcoming the attractive forces between them. As the temperature is lowered, the molecular motion becomes less vigorous, the effects of the intermolecular attractive forces become significant, and the molecules attract each other to form a liquid. In a liquid state the molecules can move around, touching each other.

When the molecular motion becomes further inactive, the molecules only vibrate around fixed centres arranged in an orderly configuration. This is a crystal. A crystal of this kind is called a *molecular crystal*, distinguished from the metallic crystal and the ionic crystal.

The way in which the molecules are arranged in a crystal is called the *crystal structure*. Figure 1.1 shows the crystal structure common to neon, argon, krypton, and xenon. In these cases a molecule is simply an atom. A unit cubic lattice contains four atoms; $1/\sqrt{2}$ times the length a of the unit cubic lattice is equal to the diameter d of a rare-gas atom. Let V be the volume of one molar crystal extrapolated to the absolute zero temperature. We then have the relation $V = Na^3/4$, where N is the Avogadro number. The values of d can be determined from this relation. Table 1.1 lists the values of V and d.

The fact that the rare-gas atoms are separated in equilibrium at the interparticle spacing d means that if one attempts to reduce the interparticle distance a strong repulsive force will be exerted between the molecules.

Based on the considerations described above, we may depict in Figure 1.2 the potential energy between two rare-gas atoms as a function of the distance r between their centres. As the atoms approach from a large distance, the potential energy $U(r)$ decreases from zero, indicating an attractive force. As they approach further within a distance representing an atomic diameter, the potential energy increases steeply, corresponding to the repulsive force. Such a potential energy is called the *intermolecular potential.*

When an electrically charged body approaches close to a light material the latter will be attracted. This is because the light material is polarized under the influence of the electric field produced by the electrically charged body. The attractive forces

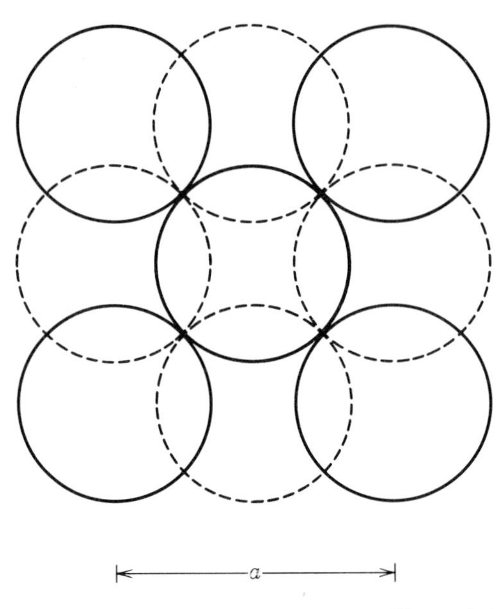

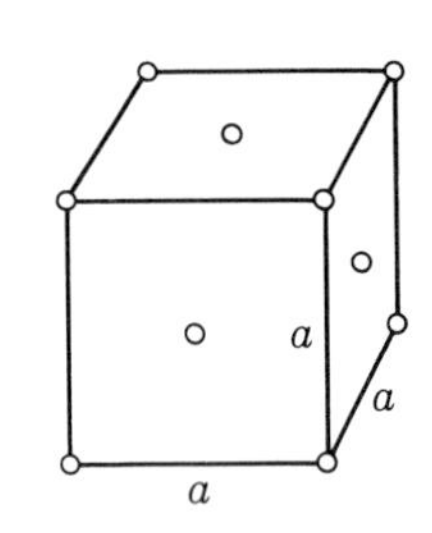

Figure 1.1

between two molecules also stem from such a phenomenon of electrical induction. Admittedly, the rare-gas atoms are neutral and are not electrically charged. Owing to electronic fluctuations, however, the distribution of the electrons may deviate in a certain direction at a given instance. Such a deviation of electric charges acts to polarize through electric induction the other rare-gas atom located nearby; those two atoms thereby attract each other.

Let us consider how the potential energy corresponding to such an attractive force changes as a function of the distance r between two rare-gas atoms. The electric field produced by the charge deviation in the first rare-gas atom at the position of the second atom is proportional to r^{-3}. The polarization of the second atom is proportional to this electric field, so that the electric field due to this polarization at the site of the first atom is proportional to $r^{-3} \times r^{-3}$. From such an

Table 1.1 Molar volume V and diameter d of rare-gas atoms

	$V(\mathrm{cm}^3)$	$d(\text{Å})$
Ne	13.06	3.13
Ar	22.3	3.74
Kr	26.8	3.98
Xe	35.5	4.37

The values of V are taken from G. A. Cook (Ed.) (1961). *Argon, Helium and the Rare Gases*, Vol. I, Interscience.

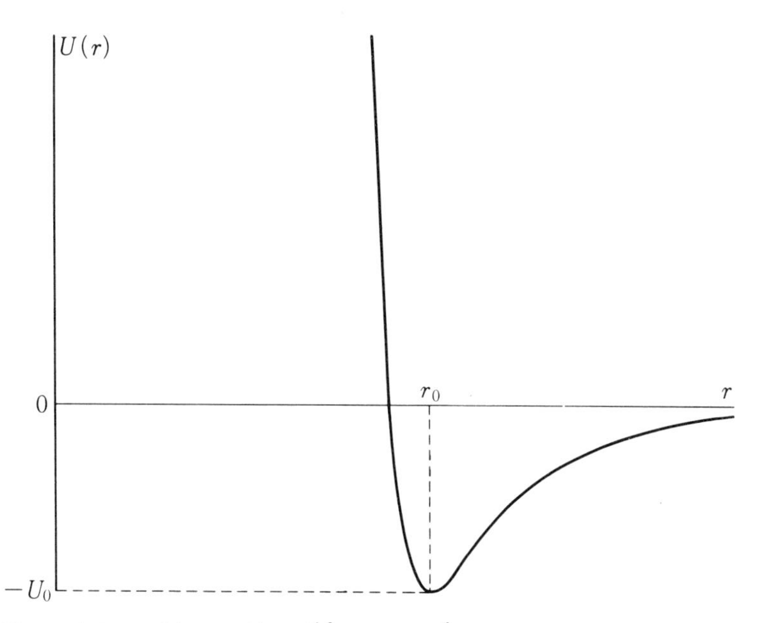

Figure 1.2 $U(r) = U_0[(r_0/r)^{12} - 2(r_0/r)^6]$ representing the potential between rare-gas atoms

argument it is clear that the attractive portion of the intermolecular potential $U(r)$ is proportional to r^{-6} at large distances. This feature has been taken into account in Figure 1.2.

The following questions now arise. (1) How accurately can one determine the intermolecular potential $U(r)$ for the rare gases, He, Ne, Ar, Kr, and Xe? (2) How can one treat the intermolecular potentials for those molecules consisting of two or more atoms? (3) How does the depth of intermolecular potentials depend on the orientation of the molecules? (4) How can one explain various structures of molecular crystals in terms of such intermolecular potentials? The answer to (1) will be given in Chapter 4, Chapter 6, and Chapter 10; (2) and (3) will be answered in Chapter 7 and Chapter 8; (4) will be answered in Chapter 9.

1.2 Chemical bond between hydrogen atoms and repulsive force between helium atoms

When two rare-gas atoms approach closer than their touching distance, a strong repulsive force is exerted between them. On the other hand, when two hydrogen atoms touch each other, these will be chemically bound to form a hydrogen molecule H_2. We now explain the remarkable difference between those two cases.

Just as light has dual character – corpuscular character as a photon and wave nature as seen in the interference and diffraction phenomena – so the electron likewise has both corpuscular and wave characters. When a flow of electrons, i.e. the electron beam, passes through a metallic foil or a crystal, it exhibits a diffraction phenomenon similar to that of the X-ray. Determining the wavelength of an electron from the crystal diffraction, we find the relation

$$p = h/\lambda$$

between the momentum p of the electron and wavelength λ, where h is the Planck constant.

In addition to such freely moving electrons, there exist electrons found in various atoms; those electrons attach individual characters to the atoms. In the hydrogen atom, an electron moves around a proton. In this case, since the electron is light and the atom small, the wavelength of the electron is comparable to the atomic dimension. (Because of this, the position and the momentum of an electron cannot be determined simultaneously in an atom; instead of Newtonian mechanics, quantum mechanics is required.)

A steady state of an electron moving around a proton implies existence of a standing wave of the electron around the proton. In a two-dimensional analogy, such a standing wave resembles vibration of a circular membrane. In the vibration of such a membrane, there exist a fundamental vibration with no nodal lines and three mutually independent vibrations with a nodal line; thus standing waves have discrete eigenmodes. A steady state of an electron in the hydrogen atom must also correspond to one of those discrete eigenmodes of the standing wave. The state of the hydrogen atom corresponding to the fundamental vibration does not have a node (a nodal surface in this case) in the standing wave of the electron. This state is called the ground state because it has the lowest energy. When the standing wave of an electron has a node, its wavelength is reduced; the momentum and thus the energy are thereby increased. These are the excited states.

The helium nucleus contains two protons in addition to neutrons. In the helium atom He, two electrons move around such a nucleus. In its ground state, the two electrons, form standing waves without nodes.

The electron by itself has an internal degree of freedom. The electron possesses an angular momentum called the spin; the component of the spin with respect to an arbitrarily chosen z axis can take on only a value $+1/2$ or $-1/2$ in units of $\hbar$ ($\hbar = \hbar/2\pi$). In the ground state of the helium atom, the total spin momentum of the two electrons forming a nodeless standing wave is zero. Generally in an atom, the number of electrons forming a standing wave is limited to one or two; when the number is two, the total spin is always zero. This is the *Pauli principle* (W. Pauli, 1924).

The lithium atom Li with atomic number 3 has three electrons. In the ground state a pair of electrons with zero total spin form a nodeless standing wave; the remaining electron forms a wave with higher energy having a nodal surface. Since this electron can be removed rather easily, the lithium has a tendency to become a singly charged positive ion.

There are four kinds of standing waves with a nodal surface in an atom: one having a spherical nodal surface with the nucleus at its centre, and three with mutually perpendicular plane nodal surfaces passing through the nucleus. At most, eight electrons can thus form standing waves with one nodal surface. Neon Ne with atomic number 10 consists of those pairs of electrons with zero total spin, forming standing waves of zero or one nodal surface.

In a hydrogen molecule two electrons form nodeless standing waves around two protons separated at 0.74 Å. In this case also the Pauli principle applies and the total spin of the two electrons is zero. The shape of a standing wave of the electrons in a hydrogen molecule is that of a spheroid; its effective wavelength is extended more than in the case of two separate hydrogen atoms. Correspondingly the kinetic energy of the electrons is reduced. This is the mechanism by which two hydrogen atoms can form a stable H_2 molecule.

When two helium atoms contact each other, four electrons surround two nuclei. Since the number of electrons capable of assuming the nodeless standing-wave state is limited to two, the remaining two electrons form a standing wave with one nodal surface. Its nodal surface is the equidistant plane between the two helium atoms. As the distance between the nuclei is reduced the effective wavelength associated with that nodal standing wave becomes short; correspondingly the kinetic energy of the electrons increases. This provides a physical explanation for the repulsive force between two helium atoms at sufficiently short distances.

1.3 The van der Waals equation of state

At pressure P and absolute temperature T, the volume V of one molar ideal gas satisfies

$$PV = RT,$$

where R is the gas constant. Since the Boltzmann constant $k = 1.3806 \times 10^{-16}$ erg/deg is obtained as R divided by the Avogadro number, this formula becomes

$$Pv = kT,$$

where v is the specific volume per molecule. Generally a formula relating P, v, and T is called the *equation of state.*

In 1873 van der Waals advanced a semi-empirical formula

$$\left(P + \frac{a}{v^2}\right)(v - b) = kT \tag{1.1}$$

as the equation of state for a real gas. Here the constants a and b are related to the intermolecular force of the gas under consideration; these are positive quantities characteristic of the gas.

This formula can be derived in the following way. Since a molecule has a finite size, the space in which gaseous molecules can move freely like those in an ideal gas is smaller than the actual volume. Taking this effect into account, we replace the v in the ideal-gas equation of state $Pv = kT$ by $v - b$. Also attractive forces are exerted between molecules at distances where the molecules do not touch each other; molecules located near the wall are thereby drawn inward. As a result the actual pressure to the wall is somewhat reduced as compared with the case of an ideal gas. The amount of such a reduction is estimated to be proportional to the product between the density of the molecules near the wall and that in the inner part. This effect thus leads us to replace P by $P + a/v^2$ in the ideal-gas equation of state $Pv = kT$.

We have thus seen that a and b in the van der Waals equation of state arise from taking into account the effects of attractive and repulsive forces between the molecules. It is in this sense that the attractive force between molecules is called the *van der Waals attractive force.*

Generally the equation of state for a gas approaches that of an ideal gas as the density decreases. Expanding the van der Waals equation of state (1.1) in a power series of $1/v$, we have

$$\frac{Pv}{kT} = 1 + \frac{1}{v}\left(b - \frac{a}{kT}\right) + \frac{b^2}{v^2} + \ldots . \tag{1.2}$$

The result of accurate measurements for actual gases can usually be expressed as

$$\frac{Pv}{kT} = 1 + \frac{B(T)}{v} + \frac{C(T)}{v^2} + \ldots , \tag{1.3}$$

where $B(T)$ and $C(T)$ are functions of temperature; these are, respectively, called the *second* and the *third virial coefficients.* The name virial coefficient has stemmed from similarity between Pv and a quantity, virial, introduced by Clausius. Incidentally, the term virial derives from a Latin word meaning force.

The second virial coefficient $B(T)$ takes on positive values at high temperatures and becomes negative at low temperatures. This can be seen from equation 1.2. At the boundary temperature, i.e. $B(T) = 0$, Boyle's law applies even in the domain of relatively high densities; this temperature is called the *Boyle temperature.* The deeper the intermolecular potential, the higher the Boyle temperature.

The virial coefficients are given as integrals involving intermolecular potentials; conversely one can determine the intermolecular potential when the virial coefficients are measured over a wide range of parameters. These aspects will be treated in Chapters 5, 6, and 7.

1.4 The van der Waals attractive force between macroscopic bodies

One may easily expect that the van der Waals attractive force, originally introduced between the molecules, may also operate between two macroscopic bodies when they are extremely close to each other. Recently such a force has been actually measured in the Soviet Union, the Netherlands, and England. We need a little preparation before going into a description of those results.

The potential $U(r)$ between spherically symmetric molecules such as Ar atoms is proportional to r^{-6} when the distance r between the centres is relatively large. This has been explained earlier; the essential point is the following. The electric field arising from fluctuations of the electron density in one of the molecules acts to polarize the other molecule; the electric field produced by such a polarization in turn acts onto the fluctuations in the first molecule. Strictly speaking, the electric

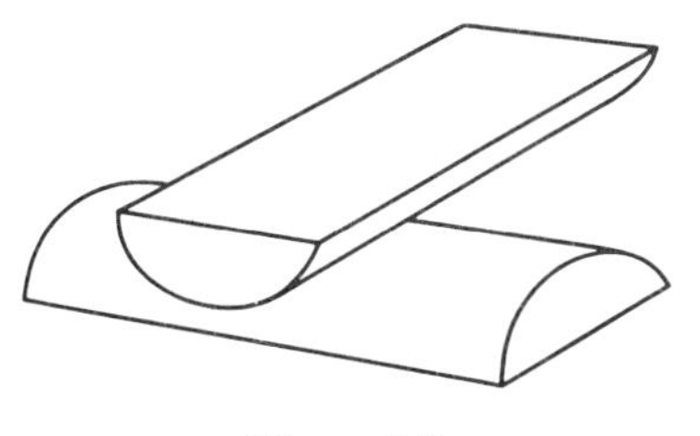

Figure 1.3

field propagates at the velocity of light in these circumstances; when two molecules are separated at a distance comparable to $1/2\pi$ times the wavelength of the light characteristic of those molecules, the effects of electric fields mentioned above cannot be treated in a quasistatic manner. In other words, when the electric field produced by fluctuations propagates from one molecule to the other and the electric field produced by polarization acts back to the original position, the state of the fluctuations may be somewhat different from the original one. As a result of such a *retardation* effect of electric fields, the intermolecular potential $U(r)$ is known theoretically to become proportional to r^{-7} at extremely large values of r.

For molecules of ordinary sizes, the attractive forces become extremely weak at such large distances, so that the retardation phenomena need not be considered. For a macroscopic body, however, such a weak force may be observed in a collective manner.

Consider two glass rods with a semicircular cross section, on the curved surfaces of which thin plates of mica are adhered; as Figure 1.3 illustrates, the mica surfaces are placed crosswise close to each other. The potential energy of the van der Waals attractive force is measured as a function of the distance D between the mica surfaces. Israelachvili and Tabor (J. N. Israelachvili and D. Tabor (1972). *Proc. Roy. Soc., London,* A. **331**, 19) obtained the following result over the range of D from 15 Å to 1300 Å: for $D < 100$ Å the energy is proportional to D^{-2}; for $D > 800$ Å it is proportional to D^{-3}. These correspond, respectively, to the integrated results of intermolecular potential proportional to r^{-6} and r^{-7} over the macroscopic bodies of mica.

Problem

Show that the potential energy of the van der Waals attraction between the two mica is inversely proportional to the square of the distance D when an attractive potential between volume elements of mica is inversely proportional to the sixth power of the distance.

Solution

We may regard the mica as semi-infinite bodies, neglecting the curvature. Choosing the axes of coordinates as shown in Figure 1.4, we calculate the potential energy per

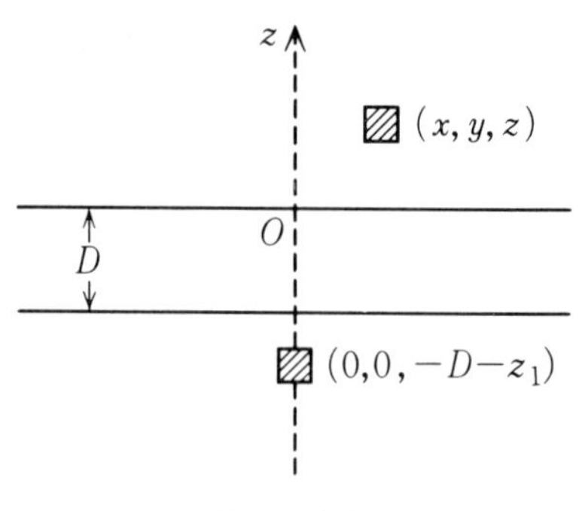

Figure 1.4

unit area of the spacing to be proportional to

$$\int_0^\infty \int_0^\infty \int_{-\infty}^\infty \int_{-\infty}^\infty \frac{dxdy}{[x^2 + y^2 + (z + z_1 + D)^2]^3} dzdz_1.$$

Choosing x/D, y/D, z/D, and z_1/D, as new integration variables, we find that this integral is proportional to D^{-2}.

The mica strongly absorbs ultraviolet with wavelengths in the vicinity of 1000 Å. When D increases to such an extent that it takes on a value close to $1/2\pi$ times such a characteristic wavelength, the retardation effects of the electric field mentioned above become significant.

In this book we treat those molecules whose dimensions are of the order of a few Angstroms; the effects of the electric-field retardation are thus negligible.

Chapter 2

The Symmetry of Molecules

2.1 Structure of molecules

The structural formula of a chemical compound may be obtained by connecting the atomic symbols by a number of bond lines which is equal to the valence of the atoms. We express carbon dioxide CO_2, water H_2O, and ethylene C_2H_4 for example, as

$$O{=}C{=}O \qquad H\diagup^{O}\diagdown H \qquad \begin{matrix} H & & H \\ & C{=}C & \\ H & & H \end{matrix}$$

These, respectively, represent examples for a linear molecule, nonlinear molecule, and a plane molecule; thus the structural formulas as expressed above exhibit symmetries of the molecules as well. For a three-dimensional molecule, however, the structural formula cannot describe the symmetry of the molecule.

Although both CO_2 and H_2O are molecules consisting of three atoms, they differ in normal modes of vibrations and rotational degrees of freedom. Generally, the normal modes of molecular vibrations are investigated through experiments of infrared absorptions and of the Raman effect. The Raman effect is a phenomenon in which the frequency of the scattered light differs from that of the light incident on the molecule by an amount corresponding to the frequency of the normal vibration at which the polarizability of the molecule changes. In particular, it is known that when a centre of symmetry exists in a molecule, the normal vibration appearing in the Raman effect does not appear in the infrared absorption. Furthermore, the moment of inertia of a molecule can be determined through investigation of the rotational spectra of the molecule by means of microwaves. Through these methods, we obtain information on the symmetry, the *bond angle*, and the *bond length* of the molecule. For example, H_2O in the gaseous state has: $\angle HOH = 104.5^\circ$ and the length of OH = 0.96 Å.

In the case of more complex molecules, the structures can be determined through experiments of electron diffraction and neutron diffraction. Table 2.1 lists some of the quantities which will be used in later chapters.

We remark on the molecules in the table that carbon disulphide S=C=S and acetylene H—C≡C—H are linear; benzene C_6H_6, hexagonal; CH_4, CF_4, and SiF_4 are of tetrahedral symmetry.

Furthermore, molecules such as boron trifluoride BF_3 and boron trichloride

Table 2.1 Interatomic distances

		bond	length (Å)
hydrogen	H_2	H–H	0.74
nitrogen	N_2	N–N	1.094
oxygen	O_2	O–O	1.10
fluorine	F_2	F–F	1.44
carbon dioxide	CO_2	C–O	1.15
carbon disulfide	CS_2	C–S	1.54
methane	CH_4	C–H	1.093
carbon tetrafluoride	CF_4	C–F	1.36
silicon tetrafluoride	SiF_4	Si–F	1.54
ethane	C_2H_6	C–C	1.533
ethylene	C_2H_4	C–C	1.33
acetylene	C_2H_2	C–C	1.20
benzene	C_6H_6	C–C	1.397

BCl_3 possess trigonal symmetry; molecules such as sulphur hexafluoride SF_6 and uranium hexafluoride UF_6 have octahedral symmetry.

In the following three sections we shall consider how those symmetries inherent in individual molecules may be explained through quantum-mechanical considerations and what are the states of electrons around those molecular structures.

2.2 The ground state of atoms

In the stationary state, the total orbital angular momentum and the total spin angular momentum of an atom are individually conserved within the nonrelativistic approximations. In classical terms, the magnitudes and the directions of those angular momenta are kept constant. Naturally the z component of an angular momentum in these circumstances can take on various values depending upon the choice of the direction of the z axis.

For a given orbital angular momentum, when the maximum possible value of its z component is $L\hbar$, this L is the number characterizing the magnitude of the orbital angular momentum. Obviously the minimum value in this case is $-L\hbar$. A fundamental property of the angular momentum is that the z component can take on discrete values,

$$L, L-1, L-2, \ldots, -L,$$

in units of $\hbar$. It then follows that $2L + 1$ must be one of the positive integers $1, 2, \ldots$ Similarly for the spin angular momentum S, $2S$ + ! must be a positive integer.

A significant difference between the orbital angular momentum L and the spin angular momentum S lies in the fact that while the values of L are limited to $0, 1, 2, \ldots,$ the values of S can be not only integers $0, 1, 2, \ldots,$ but also half-integers $1/2, 3/2, \ldots$. In particular, when L=S=0 the atom is spherically symmetric.

As is well known, the energy of a photon in the atomic spectrum is equal to the energy difference between two stationary states. It is in this sense that the stationary state or the energy level of an atom is called the *spectral term.* The spectral terms corresponding to various values of L are traditionally designated by the following capital letters:

L = 0 1 2 3 4 5 . . .
S P D F G H . . .

The $L = 0$ term is 'sharp' because it does not have fine structures; the designation S derives therefrom. The $L = 1$ term corresponds to the 'principal' part of the spectrum, whence the designation P comes. The spectral lines of the L=2 term are broadened owing to the fine structures; the D stems from diffuse. The spectrum associated with the L=3 term lies in the infrared domain; in analogy with the fundamental oscillations of sound wave, the designation of the F term derives from fundamental. It then proceeds in alphabetical order from G onwards.

As for the total spin S, the number of $2S + 1$, called the *multiplicity*, is attached to the left shoulder of the symbol for the spectral term. For instance, ^{1}S means the S term with the spin multiplicity 1; ^{2}P indicates the P term with the spin multiplicity 2.

Now, each electron in an atom may be approximately regarded as moving independently in the central force field produced by the nucleus and all the other electrons. The concept of the 'state of each electron' thus has a meaning; the state is characterized by the quantum number l representing the orbital angular momentum and the principal quantum number n, where $n = l + 1, l + 2, \ldots$.

To distinguish between the electrons with different values of l, one uses the following lower-case letters:

l = 0 1 2 3 . . .
s p d f . . .

Attaching the magnitude of the principal quantum number n to the left side of those symbols, one expresses the electrons as 1s, 2s, 2p, 3s, 3p, ...

The way in which the electrons in an atom are distributed in such 'orbitals' is called the *electron configuration*. For example, $1s^2 2s^2 2p$ represents the electron configuration in which two electrons are in the 1s orbitals, two electrons in the 2s orbitals, and one electron in a 2p orbital.

Now the maximum number of electrons which can occupy a state with given n and l is $2(2l + 1)$, for at most one electron can occupy each combination of the magnetic quantum number, $l, l - 1, \ldots, -l$, associated with the orbital angular momentum and the magnetic quantum number $\pm 1/2$ of the spin. When $2(2l + 1)$ electrons fill up all the states with given n and l, those electrons are said to form a *closed shell.* An atom consisting only of closed shells obviously belongs to ^{1}S term; it is thus spherically symmetric. In particular, the electron configurations for the ground state of rare-gas atoms are the following: He has a closed shell with two electrons in 1s state; on its 'outside', Ne has an additional closed shell of 8 electrons in 2s and 2p states; Ar, additional 8 electrons in 3s and 3p states; Kr, additional 18

Table 2.2 Electron configuration in the ground state of atoms

Li	Be	B	C	N	O	F	Ne
2s	$2s^2$	$2s^2 2p$	$2s^2 2p^2$	$2s^2 2p^3$	$2s^2 2p^4$	$2s^2 2p^5$	$2s^2 2p^6$
Na	Mg	Al	Si	P	S	Cl	Ar
3s	$3s^2$	$3s^2 3p$	$3s^2 3p^2$	$3s^2 3p^3$	$3s^2 3p^4$	$3s^2 3p^5$	$3s^2 3p^6$
^2S	^1S	^2P	^3P	^4S	^3P	^2P	^1S

electrons in 4s, 3d, and 4p states; Xe, additional 18 electrons in 5s, 4d, and 5p states; Rn, an additional closed shell of 32 electrons in 6s, 4f, 5d, and 6p states.

The ground states of the hydrogen atom and the helium atom are:

$$\text{H} \quad 1s\ {}^2S, \qquad \text{He} \quad 1s^2\ {}^1S.$$

In Table 2.2 we list the electron configurations of subsequent atoms up to argon. In this table we have omitted $1s^2$ for the atoms from Li to Ne and $1s^2 2s^2 2p^6$ for the atoms from Na to Ar.

To determine the ground-state configuration of an atom which has more than one electron with fixed (n,l) outside closed shells, we find it useful to employ Hund's law: 'The term which has the maximum total spin S and which maximizes L at this value of S has the lowest energy.'

2.3 Linear molecules

As the rare-gas atoms are in spherically symmetric ^1S states, almost all stable two-atom molecules have axially symmetric electronic states, namely, states in which the total orbital angular momentum around the molecular axis vanishes. Such a state is designated by the Greek character Σ corresponding to the S term; for the total spin S, the multiplicity $2S + 1$ is attached to the left shoulder of the term symbol.

When two hydrogen atoms in the ground state (^2S) approach each other, they may possibly assume a molecular state either $^1\Sigma$ or $^3\Sigma$. The singlet term $^1\Sigma$ is antisymmetric with respect to the spin wave function; the triplet term $^3\Sigma$ is symmetric. Consequently $^1\Sigma$ is symmetric with respect to the wave function $\psi(\mathbf{r}_1,\mathbf{r}_2)$ for the spatial coordinates $\mathbf{r}_1$ and $\mathbf{r}_2$; $^3\Sigma$ is antisymmetric. An antisymmetric wave function ψ vanishes at $\mathbf{r}_1 = \mathbf{r}_2$ and thus must have a node; hence it cannot be the ground state of the molecule. In fact the antisymmetric state cannot even constitute a molecule. In conclusion we find that the ground state of the H_2 molecule is $^1\Sigma$.

Intuitively speaking, two electrons with mutually opposite spins form a pair and jointly occupy the 1s orbital of the two-hydrogen atoms. A chemical bond of this type is called a *covalent bond* or an *electron pair bond.*

In addition to H_2, the ground states of many diatomic molecules such as N_2,F_2,Cl_2, and HCl are known to be $^1\Sigma$. Now, as for np orbitals, let us designate by np_0 and $np_\pm$, respectively, those with magnetic quantum number m_l, 0, and ±1 with respect to the molecular axis. We remark that p_0 extends along the molecular axis and $p_\pm$ extends in directions perpendicular to the axis. We thus understand that the 1s orbital of the hydrogen atom and the $3p_0$ orbital of the Cl atom form the electron pair bond in HCl and that $2p_0$ orbitals of the two F atoms form the electron pair bond in F_2. In N_2, the two p_0 orbitals and the two $p_\pm$ orbitals of the two atoms individually form electron pair bonds. It is in this sense that the valence of N is 3 and the structural formula of N_2 is written as N≡N.

Incidentally, the total spin of the electrons in a molecule also vanishes for ordinary chemical compounds, not only for diatomic molecules.

O_2 is an exceptional case in that it assumes the $^3\Sigma$ state. In this connection we remark that gaseous oxygen is paramagnetic while solid oxygen exhibits antiferromagnetism in the lattice structure at low temperatures.

The ground state of beryllium Be has a closed shell structure with electron configuration $1s^2 2s^2$. The reason for this atom being not inactive may be traced to the fact that one of the 2s electrons can easily rise to a 2p orbital and consequently the spin orientations of the two electrons become free. As we see in the example of beryllium chloride $BeCl_2$, the valence of Be is 2. The molecule Cl–Be–Cl is linear and reflectionally symmetric. This fact may be understood in terms of *hybridization* of the two orbitals 2s and 2p: constructing hybrid orbital functions

$$(s \pm p_0)/\sqrt{2}$$

out of normalized orbital functions s and p_0, we obtain an orbital extending in one direction and forming an electron pair bond with the Cl in this direction, and another orbital extending in the other direction and forming an electron pair bond with the Cl in that direction. Consequently, the valence of Be is characterized by the sp hybridization accompanied by orbital rising $2s^2 \to 2s2p$.

In the case of acetylene H–C≡C–H, a 2s electron in the C atom rises to a 2p orbital, so that the valence of C is 4. We may regard one of the hybrid orbital functions,

$$\frac{\lambda s - p_0}{\sqrt{(1+\lambda^2)}}, \quad \frac{s + \lambda p_0}{\sqrt{(1+\lambda^2)}} \quad (\lambda \text{ is an appropriate mixing ratio}),$$

made of a 2s orbital and a $2p_0$ orbital, as forming an electron pair bond with a hydrogen atom and the other with the accompanying C atom; likewise $2p_\pm$ orbitals form their own electron pair bonds. In fact, the electronic structure of an acetylene atom is axially symmetric.

In the case of carbon dioxide O=C=O, one considers hybrid orbitals $(s \pm p_0)/\sqrt{2}$ out of 2s and $2p_0$, based on the electron configuration $2s2p^3$ of a C atom; these form electron pair bonds with the O atoms in their respective directions. Furthermore, electron pair bonds are formed between the $2p_+$ orbital of the C atom and the $2p_+$ orbital of an O atom, and between the $2p_-$ orbital of C and the $2p_-$ orbital of the other O. Taking account of superposition of the states with those in which the roles of $2p_+$ and $2p_-$ are interchanged, we prove axial symmetry of the electronic structure.

Hydrogen cyanide H–C≡N and cyanogen N≡C–C≡N are additional examples of linear molecules which may be understood in a similar way.

Concerning the electron orbitals in linear molecules, those for which the quantum number of an orbital angular momentum around the molecular axis vanishes are generally called the σ orbitals; those with quantum numbers ± 1, the π orbitals. For instance, we may regard the triple bonds of acetylene as consisting of a bond with a σ orbital and two bonds with π orbitals. In Section 2.4 we shall find such a distinction to be useful.

2.4 Nonlinear molecules

The functions of np orbitals may be written as

$$R_{n\mathrm{p}}\,(r)\cos\theta$$
$$R_{n\mathrm{p}}\,(r)\sin\theta\, e^{\pm i\varphi}/\sqrt{2}$$

for each of the values 0, ± 1 of the magnetic quantum number m_l with respect to a given z axis. Here r, θ, and φ are the spherical polar coordinates which are related to x, y, and z via

$$x = r\sin\theta\cos\varphi,\quad y = r\sin\theta\sin\varphi,\quad z = r\cos\theta.$$

Since, on the other hand,

$$\cos\varphi = (e^{i\varphi} + e^{-i\varphi})/2,\quad \sin\varphi = (e^{i\varphi} - e^{-i\varphi})/2i$$

we may use the following set of real functions

$$p_x = R_{n\mathrm{p}}\,(r)\sin\theta\cos\varphi$$
$$p_y = R_{n\mathrm{p}}\,(r)\sin\theta\sin\varphi$$
$$p_z = R_{n\mathrm{p}}\,(r)\cos\theta$$

in place of those complex functions.

BF_3 has trigonal symmetry. Such a character of the boron atom may be explained in terms of the sp^2 hybridization accompanied by orbital rising $2s^2 2p \rightarrow 2s2p^2$. Choosing the xy plane on the molecular plane, we find that the hybrid orbital functions

$$\frac{1}{\sqrt{3}}s + \left(\sqrt{\frac{2}{3}}\right)p_x,\quad \frac{1}{\sqrt{3}}s + \left(\sqrt{\frac{2}{3}}\right)\left(-\frac{1}{2}p_x + \frac{\sqrt{3}}{2}p_y\right),$$
$$\frac{1}{\sqrt{3}}s + \left(\sqrt{\frac{2}{3}}\right)\left(-\frac{1}{2}p_x - \frac{\sqrt{3}}{2}p_y\right)$$

extend themselves toward the three corner directions of a regular triangle with the B atom on its centre; each orbital forms an electron pair bond with an F atom. Such a hybrid orbital is called a *trigonal orbital.* The two coefficients $1/\sqrt{3}$ and $\sqrt{(2/3)}$ are chosen in such a way that those hybrid orbital functions form an orthonormal system.

Methane CH_4 and carbon tetrafluoride CF_4 have tetrahedral symmetry. This valence of carbon C is explained in terms of sp^3 hybridization accompanied by $2s^2 2p^2 \rightarrow 2s2p^3$. The hybrid orbital functions

$$\tfrac{1}{2}(s + p_x + p_y + p_z)$$
$$\tfrac{1}{2}(s + p_x - p_y - p_z)$$
$$\tfrac{1}{2}(s - p_x + p_y - p_z)$$
$$\tfrac{1}{2}(s - p_x - p_y + p_z)$$

extend themselves toward the corners of a tetrahedron with the C atom at its centre; each orbital forms an electron pair bond with an H atom or an F atom. From the structure of the tetrahedron the bond angle $\angle$HCH is $109° \, 28'$.

The distinction between the σ orbital and the π orbital in the case of a linear molecule may also be applied to a plane molecule. Those described by symmetric orbital functions with respect to the molecular plane may be called the σ orbitals; those antisymmetric, π orbitals. The electrons in the π orbitals may sometimes be called the π electrons.

Ethylene

H   H
   C=C
H   H

is a plane molecule. Choosing the z axis in the direction perpendicular to the molecular plane, we characterize the electronic structure of these carbon atoms in terms of the trigonal orbitals made of $2s, 2p_x$, and $2p_y$, and the $2p_z$ orbital. The former are σ orbitals; the latter, π orbital. The bonds between C and H and one of the double bonds are those of the σ orbitals; the remaining double bond stems from the electron pair bond of the π orbital. Since the π electrons protrude from the molecular plane, the molecular shape is not so flat as the structural formula might suggest.

Benzene C_6H_6 is a plane molecule with hexagonal symmetry. A carbon atom has trigonal orbitals as in the case of ethylene, and forms electron pair bonds through the σ orbitals with the adjoining two carbon atoms and a hydrogen atom. The remaining six π electrons in a molecule form ring-shaped electron clouds above and below the molecular plane; they thereby form fluid-like electron-pair bonds in a steady state. Pauling treated these situations as a *resonance* between

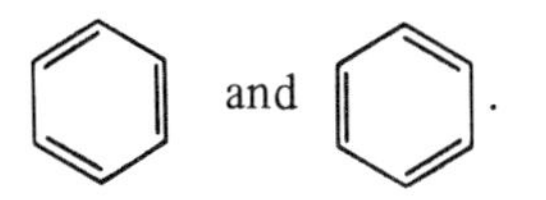

Through such a resonance the molecule acquires a higher symmetry and enhances its chemical stability.

Generally, stabilization achieved through mixing of a number of states described

by different structural formulas is called resonance. We shall find other examples of resonance in Section 3.4.

SF_6, WCl_6, UF_6, and UCl_6 have octahedral symmetry. In those molecules, the octahedral orbitals are formed through hybridization involving the d orbitals of sulphur, tungsten, and uranium.

Chapter 3

Electrostatic Properties of Molecules

3.1 Polarizability

The ratio between the dipole moment **p** induced in an atom or a molecule by an electric field and the strength **E** of the electric field, that is, the tensor α of the second rank defined through

$$\mathbf{p} = \alpha \mathbf{E},$$

is called the *polarizability*. In this section we consider the cases in which **E** is the electrostatic field. The polarizability is then characterized by three real quantities representing the principal values of the tensor α.

For rare-gas atoms or 'isometric' molecules such as CH_4 and SF_6, the principal values of α are all equal; the polarizability thus becomes a scalar quantity. Table 3.1 lists some of the measured values. We also list the values of $\alpha/(\text{radius})^3$, since the polarizability of a sphere of a perfect conductor is equal to the cubic power of its radius. Here the radii of the rare-gas atoms are taken to be half of the diameters listed in Table 1.1.

For an axisymmetric or generally 'uniaxial' molecule, the polarizability is characterized by the two components, $\alpha_{\parallel}$ and $\alpha_{\perp}$, parallel and perpendicular to the molecular axis. For those molecules without electric dipole moments, the measured values of $\alpha_{\parallel}$ and $\alpha_{\perp}$ are listed in Table 3.2.

The polarizability averaged over the molecular directions is $(\alpha_{\parallel} + 2\alpha_{\perp})/3$. The difference between $\alpha_{\parallel}$ and $\alpha_{\perp}$, that is, the anisotropy in the polarizability, is measured through depolarization of polarized lights scattered by the molecules or with the aid of the Kerr effect. (When a static electric field is applied to a gas of molecules with anisotropic polarizability, a double refraction phenomenon takes place as in the case of a single crystal with an optical axis in the direction of the electric field. This phenomenon is called the Kerr effect; the change in the principal refractive index is proportional to the square of the electric field.)

Table 3.1 Electrostatic polarizabilities α (Å^3)

	He	Ne	Ar	Kr	Xe	CH_4	CCl_4
α	0.216	0.398	1.63	2.48	4.01	2.60	10.5
$\alpha/(\text{radius})^3$		0.104	0.249	0.315	0.384		

Table 3.2 Components of electrostatic polarizabilities parallel and perpendicular to the molecular axis

	$\alpha_{\parallel}(\text{Å}^3)$	$\alpha_{\perp}(\text{Å}^3)$
H_2	0.934	0.718
N_2	2.38	1.45
O_2	2.35	1.21
Cl_2	6.60	3.62
CO_2	4.05	1.95
HC≡CH	5.12	2.43
CS_2	15.14	5.54
$(CN)_2$	7.76	3.64
C_2H_6	5.48	3.97
C_6H_6	6.35	12.31
C_6H_{12}	9.25	11.68

3.2 Electronegativity

The energy required to remove an electron from an atom in its ground state and to dissociate it into a positive ion and a free electron is called the *energy of ionization*. Its magnitudes are listed in Table 3.3 in units of electron volts (eV).

As we see in the table, it is difficult to remove electrons from those atoms which are located on the upper right side of the periodic table.

The energy liberated when a free atom and a free electron are combined to form a negative ion is called the *electron affinity*. In units of eV it has approximately the following magnitudes:

H 0.7, C 1.2, N −0.6, O 2.3, F 3.9
Si 0.6, P 0.3, S 2.5, Cl 3.7.

The electron affinity is large for halogen atoms and atoms of oxygen type.

As we observe in these examples, different atoms in a molecule have different strengths of attracting electrons: fluorine F, chlorine Cl, and oxygen O have large ionization energies and electron affinities, and consequently they have a strong tendency to attract electrons. For example, in hydrogen chloride HCl, negative charges are gathered more toward the Cl atom; positive charges move more toward the H atom. Similarly in carbon dioxide O=C=O, negative charges are attracted at both ends; positive charges gather in the central demain.

A quantitative expression for the strength of attracting electrons for a given atom in a molecule is called the *electronegativity*; usually the magnitudes in Table 3.4 determined by Pauling are used. These magnitudes have been introduced in the following way. Let $D(\text{A–B})$ be the magnitude of the energy of a simple bond between an atom A and an atom B expressed in units of kcal/mol. This quantity is equal to the dissociation energy of a bond AB. Setting the relation

$$D(\text{A–B}) = \{D(\text{A–A})D(\text{B–B})\}^{1/2} + 30(x_A - x_B)^2,$$

Table 3.3 Ionization energies of atoms (eV)

H 13.60							He 24.58
Li 5.39	Be 9.32	B 8.30	C 11.26	N 14.53	O 13.61	F 17.42	Ne 21.59
Na 5.14	Mg 7.64	Al 5.98	Si 8.15	P 10.48	S 10.36	Cl 13.01	Ar 15.75
K 4.34	Ca 6.11		Ge 7.88	As 9.81	Se 9.75	Br 11.84	Kr 14.00
Rb 4.18	Sr 5.69		Sn 7.34	Sb 8.64	Te 9.01	I 10.45	Xe 12.13
Cs 3.89	Ba 5.21			(W 7.98, U 4)			

Table 3.4 Electronegativities

H 2.1						
Li 1.0	Be 1.5	B 2.0	C 2.5	N 3.0	O 3.5	F 4.0
Na 0.9	Mg 1.2	Al 1.5	Si 1.8	P 2.1	S 2.5	Cl 3.0
			Ge 1.8	As 2.0	Se 2.4	Br 2.8
			Sn 1.8	Sb 1.9	Te 2.1	I 2.5

L. Pauling (1960). *The Nature of the Chemical Bond*, 3rd Ed, Cornell University Press.

we determine the electronegativity x_A of the atom A through applications of the formula for various combinations. Although Pauling's scale was originally introduced for a single bond, the formula may also be applied qualitatively for other kinds of bonds such as double bonds.

3.3 Dipoles, quadrupoles, and octopoles

A neutral molecule does not always mean that the charge distribution is uniformly cancelled over the entire surface of the molecule. In hydrogen chloride HCl, negative charges are found near the Cl atom; in carbon dioxide negative charges are located near the O atoms at both ends; in silicon tetrafluoride SiF_4 the F atoms attract electrons, i.e. negative charges. These, respectively, represent typical examples of a *dipole*, a *quadrupole*, and an *octopole*. We may describe the

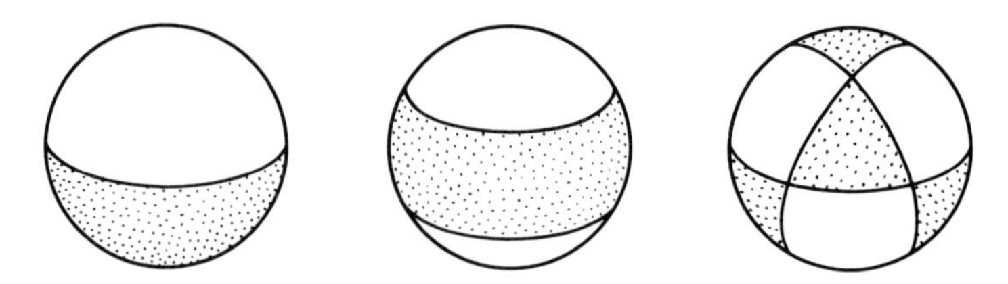

Figure 3.1 Examples of electric multipoles: from the left, a dipole, a quadrupole (unixial), and an octopole (with the symmetry of a regular tetrahedron)

characteristic features of such multipoles as in Figure 3.1, where the positive and negative charges are distinguished by light and shaded areas on the molecular surfaces.

The electric field produced at a place sufficiently far away from a dipolar molecule is determined by a vector called the *dipole moment* of the molecule. In particular, when the electric charges $\pm q$ are separated at a distance l, the magnitude of the dipole moment is defined as the product ql; it is conventionally agreed that the vector is directed from $-q$ to $+q$. The dipole moment of a molecule is measured in units of debye, which is designated by D. This unit corresponds to 10^{-18} times the cgs electrostatic unit or the Gaussian unit. For example, when $\pm e$ are separated at a distance 1 Å, the dipole moment is 4.8 D, where $e(= 4.8 \times 10^{-10}$ cgs esu) is the unit charge of an electron.

A molecule with a dipole moment is called a *polar molecule*. In the following we list some examples of dipole moments associated with polar molecules in a free gaseous state.

HF 1.83,	HCl 1.11,	HBr 0.82,	HI 0.44,
HCN 2.94,	H_2O 1.94,	H_2S 0.93,	NH_3 1.47,
PH_3 0.55	(unit D).		

The dielectric constant ϵ of a gas consisting of molecules with a dipole moment μ is given by

$$\epsilon - 1 = 4\pi n(\alpha + \mu^2/3kT). \tag{3.1}$$

Here n represents the number of molecules per unit volume, k is the Boltzmann constant, and α is the molecular dipole moment averaged over its directions. Plotting $(\epsilon - 1)/4\pi n$ on the ordinate and $1/T$ on the abscissa, one can determine the value of μ from the slope of the resultant line. For those molecules with small moments of inertia, quantum-mechanical corrections become necessary.

Problem

The potential energy of a dipole located in a uniform electric field **E** is given by $-\mu E \cos\theta$, where θ represents the angle between the two vectors. Assuming that the number of molecules oriented in a unit solid angle at angle θ is proportional to $\exp(\mu E \cos\theta / kT)$ (i.e. the Boltzmann distribution), derive equation (3.1).

Solution

The electric polarization P_0 arising from the orientation per unit volume is obviously directed along $\mathbf{E}$; its magnitude is

$$P_0 = n \int_0^{\pi} \mu \cos\theta \, e^{\mu E \cos\theta / kT} \sin\theta d\theta \Big/ \int_0^{\pi} e^{\mu E \cos\theta / kT} \sin\theta d\theta .$$

Retaining only the first-order contribution with respect to E, we have

$$P_0 = n \frac{\mu^2}{3kT} E.$$

We then calculate P as the summation of P_0 and $n\alpha E$ due to electronic polarization of molecules; it is then substituted in

$$\epsilon E = E + 4\pi P.$$

Since only the first-order effects of E are considered, we may neglect possible anisotropy of the polarizability α.

An axisymmetric quadrupole as in the case of carbon dioxide is called a *uniaxial quadrupole*; the moment Q characterizing its strength is defined as

$$Q = \frac{1}{2} \int (2z^2 - x^2 - y^2) \rho d\tau. \tag{3.2}$$

Here $\rho d\tau$ represents the electric charge contained in the infinitesimal volume $d\tau$ at the position (x, y, z), the z axis is chosen along the molecular axis, and the origin of the coordinates is taken at the centre of molecular symmetry. According to this definition the quadrupole moment of carbon dioxide takes on a negative value.

The sign of the quadrupole moment Q of the acetylene molecule HC≡CH, which has π electrons around its centre, is obviously positive. For benzene C_6H_6, we find $Q < 0$ considering the positions of the π electrons and the average positions of positive charges. For molecules like H_2, N_2, and O_2, we cannot intuitively guess the signs of the moments.

The quadrupole moments of simple molecules have been measured by Buckingham and others. In Table 3.5 we list such values together with the results of quantum-mechanical computations for H_2.

The method of measurement is schematically the following. Inside a long cylinder containing a gas, two conducting wires are stretched parallel to its axis. The electric potentials of the two wires are kept equal; a potential difference is produced between the wires and the cylindrical wall. Consider a molecule located between the two wires: its configuration becomes anisotropic, and the degree of anisotropy is proportional to Q/kT. An anisotropy of the refractive index proportional to $(\alpha_{\parallel} - \alpha_{\perp})Q/kT$ is thereby produced, where $\alpha_{\parallel}$ and $\alpha_{\perp}$ denote the

Table 3.5 Quadrupoles of molecules (10^{-26} cgs esu)

CO_2	C_6H_6	N_2	C_2H_6	O_2
−4.2	−3.6	−1.5	−0.6	−0.4
H_2	F_2	HC≡CH	Cl_2	
+0.66	+0.9	+3.0	+6.1	

The values for CO_2, N_2, C_2H_6, and O_2 are taken from A. D. Buckingham (1967), *Advances in Chemical Physics*, Vol. **12**; others are from D. E. Strogryn and A. D. Strogryn (1966), *Mol. Phys.*, **11**, 371.

polarizabilities parallel and perpendicular to the molecular axis. The quadrupole moment Q is thus determined from a measurement of the double refraction induced by such an effect.

Molecules with octopoles, such as SiF_4, and molecules with higher multipoles, such as UF_6, will be treated in Chapter 9 in connection with the intermolecular forces.

3.4 Hydrogen bond

Boric acid $B(OH)_3$ forms white, shiny, and flaky crystals. The melting point (184 ~ 186 °C) of boric acid is relatively high as compared with that of boron trifluoride BF_3 (melting point −127 °C, boiling point −101 °C, critical temperature −12.25 °C) with a similar molecular weight, which is in a gaseous state at a room temperature. This is because a plane molecule $B(OH)_3$, bound stably with three adjacent molecules on the same plane, forms a molecular layer, resulting in a flaky crystal.

In fact, the atomic configuration of two adjoining molecules may be described as:

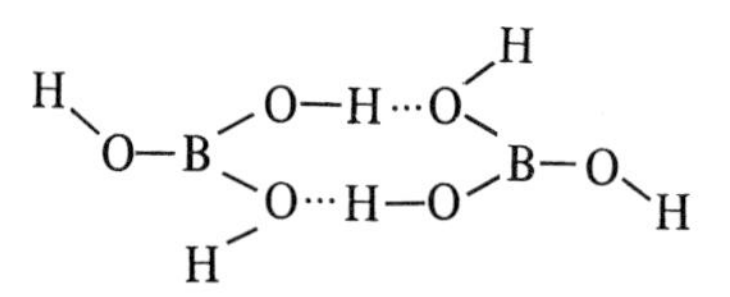

We see that each hydrogen atom is situated between two oxygen atoms.

In general, a chemical bond, stabilized by a hydrogen atom which is located between atoms of large electronegativities such as O atoms and F atoms, is called a hydrogen bond; it is represented symbolically as O–H···O. Two adjoining $B(OH)_3$ molecules are thus bound by two hydrogen bonds.

When one hydrogen atom is located between two oxygen atoms as in

$$\diagup O \;\; H \;\; O \langle ,$$

an electrostatic attractive force is exerted owing to the dipole associated with the OH radical. This energy, however, is not the essence of the hydrogen bond. The stabilizing mechanism stems rather from overlapping of those states represented by structures such as

$$\diagup O^{-}H^{+}O\diagdown \quad \text{and} \quad \diagup O^{-}H{-}O^{+}\diagdown \,.$$

Stabilization arising from mixture of states represented by various structural formulas is called the quantum-mechanical resonance (Section 2.4); the hydrogen bond is one such example.

An ordinary example of the hydrogen bond is found in water H_2O. This substance (melting point 0 °C, boiling point 100 °C, critical temperature 374.2 °C) has far higher melting and boiling points than, for example, hydrogen sulphide H_2S (melting point –85.5 °C, boiling point –60.4 °C, critical temperature 100.4 °C).

In ordinary ice, an oxygen atom is bound with four other oxygen atoms through hydrogen bonds. In these circumstances, a hydrogen atom is located closer to one of the two O atoms; on average an O atom thus has two nearby H atoms.

In this case, the H atom in O–H···O can be shifted cooperatively from one O atom to the other. When an electric field is applied such shifting takes place macroscopically; the dielectric constant of the ice thus takes on a large value (94 at –2 °C).

The hydrogen bond remains partially in the liquid water. When an electric field is applied, many molecules bound through hydrogen bonds change their directions as a whole; hence water has a large dielectric constant (81 at 18 °C).

In crystals of methanol CH_3OH or ethanol C_2H_5OH, molecules are arranged in a chain shape as (R represents an alkyl):

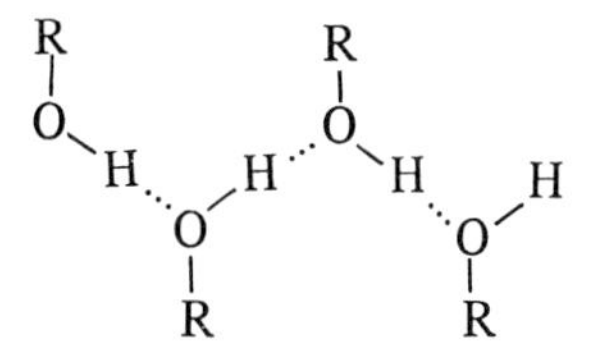

In the liquid phase, it is known that some molecules are bound in a chain or a ring shape. The fact that these alcohols dissolve well with water may be understood as resulting from hydrogen bonds commonly involved in all of those substances.

Hydrogen fluoride HF (melting point –83.7 °C, boiling point 19.5 °C, critical temperature 230.2 °C) has extremely high melting and boiling points as compared with hydrogen chloride HCl (melting point –114.2 °C, boiling point –85 °C, critical temperature 51.4 °C) and hydrogen bromide HBr; this is again due to hydrogen bonds. In solid hydrogen fluoride, the molecules are arranged in the following zigzag chain structure:

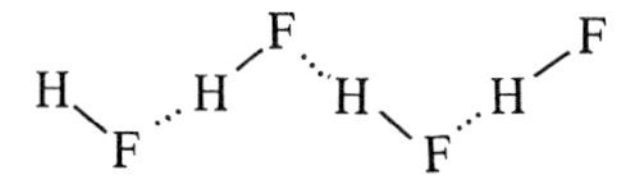

At first it might appear that a linear structure like HFHFHF would be more stable from the viewpoint of the electrostatic energy between dipoles. Taking account of the hydrogen bond F–H···F, however, arising from resonance between

$$F{-}HF\diagdown H, \qquad F^{-}H^{+}F\diagdown H, \qquad \text{and} \qquad F^{-}H{-}F^{+}\diagdown H,$$

we may understand the reason for the zigzag arrangement. Incidentally, a binding of a chain type or of a ring type exists in liquids; it is also known that polymerizations of the extent $(HF)_2 \sim (HF)_5$ at low temperatures and $(HF)_2 \sim (HF)_3$ at around 32 °C are found in the gaseous state and that hydrogen fluoride exists as HF above 90 °C.

Hereafter in this book, we treat only those simple molecules with high symmetries which do not form hydrogen bonds.

Chapter 4

The Potential of Dispersion Forces

4.1 Electric polarizability with dispersion

In this chapter it becomes necessary to consider polarizabilities with respect to a time-dependent electric field. Thus, assuming that the electric field changes in time proportionally to cos ωt, we wish to find a formula for the polarizability $\alpha(\omega)$ as a function of the angular frequency ω. This section is devoted to such a calculation.

Consider an electric field

$$E_x \cos \omega t \tag{4.1}$$

in the x direction applied to a rare-gas atom in its ground state. We may assume that the wavelength of the electric field is much longer than the atomic dimension.

We begin with the Schrödinger equation,

$$\left(H_0 + H' - i\hbar \frac{\partial}{\partial t}\right) \Psi = 0, \tag{4.2}$$

for the wave function Ψ representing the atomic state ($\hbar \equiv h/2\pi$, where h is the Planck constant). Here H_0 is the Hamiltonian in the absence of the electric field; H' represents the perturbation operator

$$H' = -p_x E_x \cos \omega t \tag{4.3}$$

arising from the electric field; and p_x denotes the operator for the x component of the dipole. We may relate the expectation value of p_x in the state Ψ with the electric field (4.1) via

$$\int \Psi^* p_x \Psi d\tau = \alpha(\omega) E_x \cos \omega t, \tag{4.4}$$

whence $\alpha(\omega)$ may be determined. The left-hand side of (4.4) means an integration over the entire domain of the variables in Ψ.

Let $\phi_0 \exp(-i\omega_0 t)$, $\phi_1 \exp(-\omega_1 t), \ldots$ be the eigenfunctions of (4.2) without perturbations, i.e.

$$H_0 \phi_n = \hbar \omega_n \phi_n, \quad n = 0, 1, \ldots, \tag{4.5}$$

where $n = 0$ corresponds to the ground state. Since the perturbation is small, Ψ may be expressed as a mixture of $\phi_0 \exp(-i\omega_0 t)$ and some other eigenfunctions:

$$\Psi = \phi_0 \exp(-i\omega_0 t) + \sum_n \phi_n \exp(-i\omega_n t) a_n(t).$$

For eigenfunctions with continuous eigenvalues, we may replace the summation by an integration. To determine the coefficients $a_n(t)$, we note from (4.2)

$$\left(H_0 - i\hbar \frac{\partial}{\partial t}\right) \sum_n \phi_n \exp(-i\omega_n t) a_n(t) = -H'\phi_0 \exp(-i\omega_0 t).$$

With the aid of (4.5) we then have

$$-i\hbar \sum_n \phi_n \exp(-i\omega_{n0} t) \frac{\partial}{\partial t} a_n(t) = p_x \phi_0 E_x \frac{e^{i\omega t} + e^{-i\omega t}}{2},$$

where $\omega_{n0} \equiv \omega_n - \omega_0$. Multiplying this by ϕ_n^*, carrying out integrations with respect to the variables, and evoking the conditions for orthonormality

$$\int \phi_n^* \phi_m \, d\tau = \begin{cases} 1 & n = m \\ 0 & n \neq m \end{cases},$$

we find

$$-i\hbar \exp(-i\omega_{n0} t) \frac{\partial}{\partial t} a_n(t) = \langle n \mid p_x \mid 0 \rangle E_x \frac{e^{i\omega t} + e^{-i\omega t}}{2}.$$

Here, in general, $\langle n \mid p_x \mid m \rangle$ is the matrix element defined as

$$\int \phi_n^* p_x \phi_m d\tau.$$

A periodic solution for the foregoing equation is

$$\exp(-i\omega_{n0} t) a_n(t) = \frac{1}{2\hbar} \langle n \mid p_x \mid 0 \rangle E_x \left(\frac{e^{i\omega t}}{\omega_{n0} + \omega} + \frac{e^{-i\omega t}}{\omega_{n0} - \omega} \right).$$

The coefficients in the expression of Ψ are thereby determined. Since $\langle 0 \mid p_x \mid 0 \rangle = 0, a_0(t)$ is zero.

Substituting this Ψ into (4.4), we may calculate the expectation value of p_x in the following way:

$$\int \Psi^* p_x \Psi d\tau = \sum_n \exp(i\omega_{n0} t) a_n^*(t) \langle n \mid p_x \mid 0 \rangle + \sum_n \langle 0 \mid p_x \mid n \rangle \exp(-i\omega_{n0} t) a_n(t)$$

$$= \frac{2}{\hbar} \sum_n \frac{|\langle 0 \mid p_x \mid n \rangle|^2 \omega_{n0}}{\omega_{n0}^2 - \omega^2} E_x \cos \omega t.$$

(Here we used the general relationship that $\langle m \mid f \mid n \rangle$ is complex conjugate to $\langle n \mid f \mid m \rangle$ *for an arbitary operator* f representing a physical quantity.) Comparison of this with (4.4) yields

$$\alpha(\omega) = \frac{2}{\hbar} \sum_n \frac{|\langle 0 | p_x | n \rangle|^2 \omega_{n0}}{\omega_{n0}^2 - \omega^2} . \tag{4.6}$$

For contributions from those states with continuous eigenvalues, the summation is to be replaced by an integration.

Light passing through a prism is resolved into its spectral components, since the refractive index of the glass depends on the frequency of the light. Equation (4.6) essentially describes the same effect; it thus represents the *dispersion* of the polarizability. The nomenclature, *dispersion force* between molecules, derives from such an expression, as we shall see in the following sections.

4.2 Time-independent perturbation

In the previous section we treated a perturbation with periodic variation in time. In this section we consider a time-independent perturbation as a preparation for the following three sections.

A time-independent wave function ψ satisfies the Schrödinger equation,

$$(H_0 + J - W)\psi = 0, \tag{4.7}$$

where J represents a perturbation operator and W denotes the energy eigenvalue. It is assumed that the eigenvalues $\hbar\omega_n$ and the eigenfunctions ϕ_n for the perturbed Hamiltonian H_0 are known:

$$H_0 \phi_n = \hbar\omega_n \phi_n, \quad \nabla n = 0, 1, \ldots .$$

Expanding an eigenfunction ψ as

$$\psi = \sum_n c_n \phi_n,$$

we find that (4.7) reduces to

$$\sum_n c_n (\hbar\omega_n - W + J)\phi_n = 0.$$

Multiplying ϕ_m^* by this and carrying out integration with the aid of the orthonormality conditions for the eigenfunctions, we obtain

$$(W - \hbar\omega_m)c_m = \sum_n c_n \langle m | J | n \rangle , \tag{4.8}$$

where $\langle m | J | n \rangle$ is the matrix element defined as

$$\int \phi_m^* J \phi_n d\tau .$$

It will become necessary in later sections to find a formula describing how the energy eigenvalue of the ground state $n = 0$ may change due to a perturbation. Without perturbations,

$$c_0 = 1, \quad c_n = 0 \; (n \neq 0), \quad W = \hbar\omega_0 .$$

In the presence of a perturbation, these are changed so that

$$c_0 = 1 + c_0^{(2)} + \ldots$$
$$c_n = c_n^{(1)} + \ldots \qquad (n \neq 0)$$
$$W = \hbar\omega_0 + W_1 + W_2 + \ldots .$$

Here W_1, and $c_n^{(1)}$ are linearly proportional to J, while W_2 and $c_0^{(2)}$ are proportional to its square.

Taking the first-order terms in J, we obtain from (4.8)

$$W_1 = \langle 0 \mid J \mid 0 \rangle. \tag{4.9}$$

That is, the first-order perturbation in the energy is equal to the expectation value of the perturbation operator J in the unperturbed state. The coefficients of expansion for the wave function are

$$c_n^{(1)} = -\frac{\langle n \mid J \mid 0 \rangle}{\hbar\omega_{n0}} \qquad (\omega_{n0} \equiv \omega_n - \omega_0).$$

With the aid of these, the second-order perturbation in energy may be obtained from (4.8):

$$W_2 = -\sum_n{}' \frac{\langle 0 \mid J \mid n \rangle \langle n \mid J \mid 0 \rangle}{\hbar\omega_{n0}}, \tag{4.10}$$

where Σ' means omission of the term with $n = 0$ in the summation. When there is a contribution from states with continuous eigenvalues, that part of the summation is to be replaced by an appropriate integration.

The third-order perturbations are somewhat complicated. In particular when $\langle 0 \mid J \mid 0 \rangle = 0$, i.e. when the first-order perturbation vanishes, we have

$$W_3 = \sum_n{}' \sum_m{}' \frac{\langle 0 \mid J \mid n \rangle \langle n \mid J \mid m \rangle \langle m \mid J \mid 0 \rangle}{\hbar^2 \omega_{n0}\omega_{m0}}. \tag{4.11}$$

4.3 Attractive potential between two molecules

Let us calculate the interaction potential between two rare-gas atoms in the ground state separated at a large distance. To solve this problem we use perturbation theory, that is, we regard two isolated atoms as an unperturbed system and treat the energy of electric interaction as a perturbation operator. Such an electric interaction may be expanded at a large distance r between the two atoms in terms of the dipole–dipole interaction proportional to r^{-3}, the dipole–quadrupole interaction proportional to r^{-4}, and so on.

It is important to note that those dipoles, etc., do not correspond to those electrostatic dipoles, etc., inherent in the molecules: those are operators. Since the atom under consideration is spherically symmetric, the expectation values of the dipole, quadrupole, etc. are zero. In other words, no contributions stem from the first-order perturbation.

The second-order perturbation produces finite results. It is proportional to r^{-6} for the dipole–dipole interaction and to r^{-8} for the dipole–quadrupole interaction. In the following, we consider the most important case of the interaction of a dipolar type only; let J be the energy operator of such an interaction. Denoting the dipoles of atom 1 and atom 2 as $\mathbf{p}_1$ and $\mathbf{p}_2$, we have

$$J = r^{-3}\,(p_{1x}p_{2x} + p_{1y}p_{2y} - 2p_{1z}p_{2z}), \tag{4.12}$$

where the z axis is chosen in the direction connecting the two atoms.

We write the energy difference between the ground state 0 and an excited state ρ as $\hbar\omega_\rho$ for atom 1 and use the subscript σ for atom 2. The second-order perturbation energy $W_2(r)$ for such a diatomic system is expressed in accord with (4.10) as

$$W_2(r) = -\sum_{\rho,\sigma\neq 0}\sum \frac{\langle 00 \mid J \mid \rho\sigma\rangle\langle\rho\sigma \mid J \mid 00\rangle}{\hbar\omega_\rho + \hbar\omega_\sigma}, \tag{4.13}$$

where $\langle 00 \mid J \mid \rho\sigma\rangle$ represents the matrix element between the two eigenstates $(0,0)$ and (ρ,σ) for the unperturbed system. It is clear from the form of (4.13) that $W_2(r)$ is negative and proportional to r^{-6}, i.e.

$$W_2(r) = -\mu_{12}r^{-6}. \tag{4.14}$$

For convenience in calculations, we write (4.12) as

$$J = r^{-3}\mathbf{p}_1 \cdot \mathbf{T} \cdot \mathbf{p}_2, \quad \mathbf{T} = \mathbf{1} - 3\mathbf{e}_{12}\mathbf{e}_{12}.$$

Here $\mathbf{e}_{12}$ is a unit vector directing from atom 1 to atom 2; $\mathbf{1}$ represents the unit tensor. Substitution of this expression into (4.13) yields

$$\mu_{12} = \sum_{\rho,\sigma\neq 0}\sum \frac{\langle 0 \mid \mathbf{p}_1 \mid \rho\rangle . \mathbf{T}. \langle 0 \mid \mathbf{p}_2 \mid \sigma\rangle\langle\sigma \mid \mathbf{p}_2 \mid 0\rangle . \mathbf{T}. \langle\rho \mid \mathbf{p}_1 \mid 0\rangle}{\hbar\omega_\rho + \hbar\omega_\sigma}.$$

This formula may be transformed, with the aid of the symbol tr meaning the diagonal sum (trace) of a tensor, into the form

$$\mu_{12} = \sum_{\rho,\sigma\neq 0}\sum \frac{|\langle 0 \mid p_{1x} \mid \rho\rangle|^2\, |\langle 0 \mid p_{2x} \mid \sigma\rangle|^2}{\hbar\omega_\rho + \hbar\omega_\sigma}\ \mathrm{tr}\, \mathbf{T}\,.\,\mathbf{T}\ .$$

Here we used the fact that the $\hbar\omega_\rho$ in the denominator is independent of the magnetic quantum numbers for the orbital angular momentum of the atom, and the tensor $\Sigma' \langle 0 \mid \mathbf{p}_1 \mid \rho\rangle\langle\rho \mid \mathbf{p}_1 \mid 0\rangle$, summed over the states distinguished by these magnetic quantum numbers, is spherically symmetric and equal to $\Sigma' \langle 0 \mid p_{1x} \mid \rho\rangle\langle\rho \mid p_{1x} \mid 0\rangle\, \mathbf{1}$. Another way of expressing it is 'the tensor $\Sigma' \langle 0 \mid \mathbf{p}_1 \mid \rho\rangle\langle\rho \mid \mathbf{p}_1 \mid 0\rangle$, summed over an orthonormal set which spans an irreducible representation space of the rotation group, is rotation invariant . . . .'

Problem

Consider an electron in a spherically symmetric field. For the orbital angular momentum $l = 1$, the basis for the $(2l + 1)$-dimensional representation of the rotation group may be taken to be

$$xR_1(r)/r, \quad yR_1(r)/r, \quad \text{and} \quad zR_1(r)/r,$$

which have been used in Section 2.4. Prove then

$$\Sigma' \langle 0 \mid \mathbf{p} \mid \rho \rangle \langle \rho \mid \mathbf{p} \mid 0 \rangle = \Sigma' \langle 0 \mid p_x \mid \rho \rangle \langle \rho \mid p_x \mid 0 \rangle \mathbf{1}.$$

Solution

Setting

$$\iiint R_0(r)\mathbf{p}xR_1(r)r^{-1}\,dxdydz \equiv \langle 0 \mid \mathbf{p} \mid x \rangle$$

where $R_0(r)$ is the wave function for the ground state, we find that the left-hand side of the equation to be proved is

$$\langle 0 \mid \mathbf{p} \mid x \rangle \langle x \mid \mathbf{p} \mid 0 \rangle + \langle 0 \mid \mathbf{p} \mid y \rangle \langle y \mid \mathbf{p} \mid 0 \rangle + \langle 0 \mid \mathbf{p} \mid z \rangle \langle z \mid \mathbf{p} \mid 0 \rangle.$$

Since the product of (x,y,z) and the electronic charge produces (p_x,p_y,p_z), we obtain the solution.

Furthermore,

$$\begin{aligned} \text{tr}\,\mathbf{T}\,.\,\mathbf{T} &= \text{tr}(\mathbf{1} - 3\mathbf{e}_{12}\mathbf{e}_{12})\,.\,(\mathbf{1} - 3\mathbf{e}_{12}\mathbf{e}_{12}) \\ &= \text{tr}(\mathbf{1} + 3\mathbf{e}_{12}\mathbf{e}_{12}) \\ &= 6. \end{aligned}$$

Consequently we obtain (F. London 1930)

$$\mu_{12} = 6 \sum_{\rho,\sigma \neq 0}\sum \frac{|\langle 0 \mid p_{1x} \mid \rho \rangle|^2 \, |\langle 0 \mid p_{2x} \mid \sigma \rangle|^2}{\hbar\omega_\rho + \hbar\omega_\sigma}. \tag{4.15}$$

Here again for the contributions from those states with continuous eigenvalues, the summation is to be replaced by an integration.

With the aid of an identity,

$$\frac{1}{a+b} = \frac{2}{\pi}\int_0^\infty \frac{ab}{(a^2+u^2)(b^2+u^2)}\,du, \qquad (a, b > 0)$$

obtainable elementarily or with the aid of an integration on the complex plane, (4.15) may be rewritten in the form,

$$\mu_{12} = \frac{3\hbar}{\pi}\int_0^\infty \alpha_1(\mathrm{i}\omega)\alpha_2(\mathrm{i}\omega)d\omega, \tag{4.16}$$

where

$$\alpha_1(i\omega) = \frac{2}{\hbar} \sum_{\rho \neq 0} \frac{|\langle 0 | p_{1x} | \rho \rangle|^2 \omega_\rho}{\omega_\rho^{\ 2} + \omega^2} . \tag{4.17}$$

One obtains $\alpha_2(i\omega)$ similarly. These are functions analogous to the dispersion (4.6) of the polarizability obtained in Section 4.1. In other words, the right-hand side of (4.16) corresponds to integration of $\alpha_1\alpha_2$ along the imaginary axis on the complex ω plane.

For those atoms with accurately measured refractive indices, the dispersion of the polarizabilities may be approximated by a summation of finite terms

$$\alpha(\omega) = \frac{e^2}{m} \sum_n \frac{f_n}{\omega_n^{\ 2} - \omega^2} , \tag{4.18}$$

where e is the unit electric charge and m is the mass of an electron. The oscillator strengths f_n and the characteristic frequencies ω_n may be numerically determined by fitting this function with observational data. With the aid of the function

$$\alpha(i\omega) = \frac{e^2}{m} \sum_n \frac{f_n}{\omega_n^{\ 2} + \omega^2} \tag{4.19}$$

thus determined, Dalgarno *et al.* (1967) computed μ_{12} between rare-gas atoms; the results shown in Table 4.1.

In Tables 4.1 and 4.2, atomic units are used, in which the mass, the electric charge, and the angular momentum are measured in units of the electronic mass m, the elementary electric charge e, and $\hbar$, respectively. In these cases, the unit of length is the Bohr radius, 0.52917 Å, and the unit of energy is 27.210 eV (twice the ionization energy of the hydrogen atom).

For real values of the frequency ω, (4.18) is not a monotonous function. However, $\alpha(i\omega)$ appearing in the integrand decreases monotonously from $\alpha(0)$ to $\alpha(i\infty) = 0$, as (4.19) shows. Taking this into account, we may approximate $\alpha(i\omega)$ by

Table 4.1 Two-body coefficients μ_{12} (in atomic units)

	He	Ne	Ar	Kr	Xe
He	1.47	3.0	9.6	13	19
Ne		6.3	20	27	38
Ar			65	91	130
Kr				130	190
Xe					270

From A. Dalgarno (1967). *Advances in Chemical Physics*, Vol. 12. ($1 = 9.57 \times 10^{-13}$ erg Å^6.)

Table 4.2 Electrostatic polarizabilities α, the characteristic frequencies $\tilde{\omega}$, and ionization energies I (all in atomic units)

	α	$\hbar\tilde{\omega}$	I
He	1.38	1.03	0.903
Ne	2.66	1.19	0.793
Ar	11.08	0.71	0.579
Kr	16.73	0.61	0.515
Xe	27.29	0.49	0.446

the following single term:

$$\alpha_1(i\omega) = \alpha_1 \tilde{\omega}_1^2/(\tilde{\omega}_1^2 + \omega^2),$$
$$\alpha_2(i\omega) = \alpha_2 \tilde{\omega}_2^2/(\tilde{\omega}_2^2 + \omega^2), \tag{4.20}$$

where α_1 and α_2 are the electrostatic polarizabilities $\alpha_1(0)$ and $\alpha_2(0)$. The integration in (4.16) can then be performed as

$$\mu_{12} = \frac{3}{2}\frac{\hbar\tilde{\omega}_1\tilde{\omega}_2}{\tilde{\omega}_1 + \tilde{\omega}_2}\alpha_1\alpha_2. \tag{4.21}$$

From this we obtain a relation which expresses μ_{12} between different atoms in terms of μ_{11} and μ_{22} between similar atoms:

$$\frac{\alpha_1\alpha_2}{\mu_{12}} = \frac{1}{2}\left(\frac{\alpha_1^2}{\mu_{11}} + \frac{\alpha_2^2}{\mu_{22}}\right). \tag{4.22}$$

To a good approximation, we may also use

$$\mu_{12} = (\mu_{11}\mu_{22})^{1/2}. \tag{4.23}$$

The $\tilde{\omega}$ contained in (4.20) or in (4.21) is a frequency characterizing the properties of the atom; $\hbar\tilde{\omega}$ represents 'a kind of averaged value for the excitation energies' including the continuous eigenvalues. In Table 4.2 we list $\hbar\tilde{\omega}$ obtained from the relation $\mu = (3/4\hbar\tilde{\omega}\alpha^2$ with the aid of the values of μ in Table 4.1. The fact that $\hbar\tilde{\omega}$ is slightly larger than the ionization energy implies that the contribution from the continuous eigenvalues is not negligible.

The derivation of (4.16) by the quantum-mechanical perturbation theory is quite different from the intuitive derivation described in Section 1.1. Here let us briefly consider the relationship between the two, in connection with the treatment of polarization of an atom in the electric field produced by electric-charge fluctuations in another atom.

We choose the z axis along the line connecting atom 1 and atom 2; the x and y axes are chosen perpendicular to it. The mean square value $\langle p_{1x}{}^2 \rangle$ of the x

component of the dipole $\mathbf{p}_1$ produced in the atom 1 due to the charge fluctuations can be expressed in terms of the spectral components as

$$\langle p_{1x}^2 \rangle = \int_{-\infty}^{\infty} (p_{1x}^2)_\omega d\omega.$$

Similar expressions apply for the other components. Writing the attractive potential as $-\mu_{12}r^{-6}$, we have

$$\mu_{12} = \frac{1}{2} \int_{-\infty}^{\infty} [\alpha_2(\omega)\{(p_{1x}^2)_\omega + (p_{1y}^2)_\omega + 4(p_{1z}^2)_\omega\}$$

$$+ \alpha_1(\omega)\{(p_{2x}^2)_\omega + (p_{2y}^2)_\omega + 4(p_{2z}^2)_\omega\}] d\omega,$$

or taking account of spherical symmetry of the atom, we have

$$\mu_{12} = 3 \int_{-\infty}^{\infty} [\alpha_2(\omega)(p_{1x}^2)_\omega + \alpha_1(\omega)(p_{2x}^2)_\omega] d\omega.$$

Here $\alpha_1(\omega)$ and $\alpha_2(\omega)$ are the polarizabilities of the two atoms; their real and imaginary parts, $\alpha'(\omega)$ and $\alpha''(\omega)$, have the properties $\alpha'(-\omega) = \alpha'(\omega)$ and $\alpha''(-\omega) = -\alpha''(\omega)$.

Now, the spectral density $(p_x{}^2)_\omega$ of the fluctuations and the imaginary part of the polarizability are related to each other via

$$(p_x^2)_\omega = \begin{cases} \alpha''(\omega)\hbar/2\pi & (\omega > 0) \\ -\alpha''(\omega)\hbar/2\pi & (\omega < 0) \end{cases}$$

(H. B. Callen and T. A. Welton 1951). With the aid of this relationship and $\alpha''(\omega) = [\alpha(\omega) - \alpha(-\omega)/2\mathrm{i}$, we have

$$\mu_{12} = \frac{3\hbar}{2\pi \mathrm{i}} \left[\int_0^{\infty} \alpha_1(\omega)\alpha_2(\omega) d\omega - \int_{-\infty}^{0} \alpha_1(\omega)\alpha_2(\omega) d\omega \right].$$

Since $\alpha(\omega)$ is regular in the upper half of the complex ω plane, we may deform the contour of integration and obtain

$$\mu_{12} = \frac{3\hbar}{\pi} \int_0^{\infty} \alpha_1(\mathrm{i}\omega)\alpha_2(\mathrm{i}\omega) d\omega.$$

4.4 Three-molecular potential

The potential of the dispersion force between three rare-gas atoms may be regarded as a summation of two-body potentials. In fact the energy between three atoms due to the second-order perturbation is expressed in the form,

$$W_2(r_{12}, r_{23}, r_{31}) = -\mu_{12}r_{12}{}^{-6} - \mu_{23}r_{23}{}^{-6} - \mu_{31}r_{31}{}^{-6}, \tag{4.24}$$

where r_{12} is the distance between atom 1 and atom 2, etc. Nonadditivity of the potential of the dispersion force appears in the third-order perturbations.

For the positions $\mathbf{r}_1$, $\mathbf{r}_2$, and $\mathbf{r}_3$ of the three spherically symmetric atoms, we set

$$\mathbf{r}_j - \mathbf{r}_i = r_{ij}\mathbf{e}_{ij}, \quad |\mathbf{e}_{ij}| = 1.$$

The interaction energy J may be split into three terms:

$$J = J(12) + J(23) + J(31).$$

Here $J(12) = r_{12}^{-3}\mathbf{p}_1 \,.\, \mathsf{T}(12) \,.\, \mathbf{p}_2$, $\mathsf{T}(12) = \mathbf{1} - 3\mathbf{e}_{12}\mathbf{e}_{12}$, and similar formulas apply for $J(23)$ and $J(31)$. Using the subscripts ρ, σ, τ for the excited states of the atoms 1, 2, 3, and writing their energy differences from the ground state 0 as $\hbar\omega_\rho$, $\hbar\omega_\sigma$, $\hbar\omega_\tau$, we may express the right-hand side of (4.11) as a summation of six terms of the form,

$$\sum\sum\sum_{\rho,\sigma,\tau\neq 0} \frac{\langle 000 \mid J(12) \mid \rho\sigma 0 \rangle \langle \rho\sigma 0 \mid J(23) \mid \rho 0 \tau \rangle \langle \rho 0 \tau \mid J(31) \mid 000 \rangle}{(\hbar\omega_\rho + \hbar\omega_\sigma)(\hbar\omega_\rho + \hbar\omega_\tau)}.$$

As we have done in the previous section, we may rewrite such a term as

$$\sum\sum\sum_{\rho,\sigma,\tau\neq 0} \frac{|\langle 0 \mid p_{1x} \mid \rho \rangle|^2 |\langle 0 \mid p_{2x} \mid \sigma \rangle|^2 |\langle 0 \mid p_{3x} \mid \tau \rangle|^2}{(\hbar\omega_\rho + \hbar\omega_\sigma)(\hbar\omega_\rho + \hbar\omega_\tau)}$$
$$\times (r_{12}r_{23}r_{31})^{-3} \operatorname{tr} \mathsf{T}(12) \,.\, \mathsf{T}(23) \,.\, \mathsf{T}(31).$$

Its diagonal sum may be calculated as

$$\begin{aligned} \operatorname{tr} \mathsf{T}(12) \,.\, \mathsf{T}(23) \,.\, \mathsf{T}(31) &= \operatorname{tr}[(\mathbf{1} - 3\mathbf{e}_{12}\mathbf{e}_{12}) \,.\, (\mathbf{1} - 3\mathbf{e}_{23}\mathbf{e}_{23}) \,.\, (\mathbf{1} - 3\mathbf{e}_{31}\mathbf{e}_{31})] \\ &= 3[-2 + 3\{(\mathbf{e}_{31} \,.\, \mathbf{e}_{12})^2 + (\mathbf{e}_{12} \,.\, \mathbf{e}_{23})^2 + (\mathbf{e}_{23} \,.\, \mathbf{e}_{31})^2\} \\ &\quad -9(\mathbf{e}_{12} \,.\, \mathbf{e}_{23})(\mathbf{e}_{23} \,.\, \mathbf{e}_{31})(\mathbf{e}_{31} \,.\, \mathbf{e}_{12})] \\ &= 3(3 \cos\theta_1 \cos\theta_2 \cos\theta_3 + 1), \end{aligned}$$

where θ_1, θ_2, and θ_3 are the corner angles of the triangle formed by the three atoms. For example,

$$\cos\theta_1 = -\mathbf{e}_{31} \,.\, \mathbf{e}_{12}.$$

We have also used the formula,

$$\cos^2\theta_1 + \cos^2\theta_2 + \cos^2\theta_3 = 1 - 2\cos\theta_1 \cos\theta_2 \cos\theta_3.$$

Consequently the third-order perturbation energy is

$$W_3(r_{12}, r_{23}, r_{31}) = \nu_{123}(r_{12}r_{23}r_{31})^{-3}(3\cos\theta_1 \cos\theta_2 \cos\theta_3 + 1), \tag{4.25}$$

$$\begin{aligned} \nu_{123} = 6 \sum\sum\sum_{\rho,\sigma,\tau\neq 0} &[(\hbar\omega_\rho + \hbar\omega_\sigma)^{-1}(\hbar\omega_\rho + \hbar\omega_\tau)^{-1} \\ &+ (\hbar\omega_\rho + \hbar\omega_\sigma)^{-1}(\hbar\omega_\sigma + \hbar\omega_\tau)^{-1} \\ &+ (\hbar\omega_\rho + \hbar\omega_\tau)^{-1}(\hbar\omega_\sigma + \hbar\omega_\tau)^{-1}] \\ &\times |\langle 0 \mid p_{1x} \mid \rho \rangle|^2 |\langle 0 \mid p_{2x} \mid \sigma \rangle|^2 |\langle 0 \mid p_{3x} \mid \tau \rangle|^2, \end{aligned}$$

Table 4.3 Three-body coefficients ν (in atomic units)

	ν	$3\alpha\mu/4$
He	1.49	1.52
Ne	11.8	12.6
Ar	521	540
Kr	1560	1630
Xe	5430	5530

$(1 = 1.42 \times 10^{-13}$ erg Å$^9)$

or

$$\nu_{123} = 12 \sum\sum\sum_{\rho,\sigma,\tau\neq 0} \frac{\hbar\omega_\rho + \hbar\omega_\sigma + \hbar\omega_\tau}{(\hbar\omega_\rho + \hbar\omega_\sigma)(\hbar\omega_\sigma + \hbar\omega_\tau)(\hbar\omega_\tau + \hbar\omega_\rho)} \tag{4.26}$$

$$\times |\langle 0 | p_{1x} | \rho\rangle|^2 |\langle 0 | p_{2x} | \sigma\rangle|^2 |\langle 0 | p_{3x} | \tau\rangle|^2.$$

The nonadditive correction (4.25) is positive and produces a repulsive force when all of the θ_i are smaller than 117°; it becomes negative and produces an attractive force when one of the θ_i is greater than 126°.

As in the previous section, (4.26) can be related to an integrated quantity of a product of polarizabilities along the imaginary axis of the frequency:

$$\nu_{123} = \frac{3\hbar}{\pi} \int_0^\infty \alpha_1(i\omega)\alpha_2(i\omega)\alpha_3(i\omega)d\omega. \tag{4.27}$$

The values computed by Dalgarno *et al.* between similar atoms with the aid of the experimental formula (4.18) are shown in Table 4.3.

Within the single-term approximation as in (4.20), (4.27) reduces to

$$\nu_{123} = \frac{3}{2} \frac{\hbar\tilde{\omega}_1\tilde{\omega}_2\tilde{\omega}_3(\tilde{\omega}_1 + \tilde{\omega}_2 + \tilde{\omega}_3)}{(\tilde{\omega}_1 + \tilde{\omega}_2)(\tilde{\omega}_1 + \tilde{\omega}_3)(\tilde{\omega}_2 + \tilde{\omega}_3)} \alpha_1\alpha_2\alpha_3. \tag{4.28}$$

Combining this with (4.21) we obtain a formula relating ν and μ for similar atoms and the electrostatic polarizability α (Y. Midzuno and T. Kihara (1956), *J. Phys. Soc., Japan,* **11**, 1045):

$$4\nu = 3\alpha\mu. \tag{4.29}$$

For comparison we also list the values of $3\alpha\mu/4$ in Table 4.3.

To see the degree of nonadditivity in the potential of the dispersion force, let us consider the case in which three rare-gas atoms of the same kind are arranged in a regular triangular form in contacting with each other. Regarding the molecule as a sphere of diameter d, we obtain from (4.24)

$$W_2 = -3\mu d^{-6},$$

and from (4.25)

$$W_3 = (11/8)\nu d^{-9}.$$

Combination of these and (4.29) yields

$$\frac{W_3}{|W_2|} = \frac{11}{32}\frac{\alpha}{d^3}.$$

With the aid of Table 3.1, we obtain

Ne 0.004, Ar 0.011, Kr 0.014, Xe 0.017.

As far as the potential of the dispersion force is concerned, its nonadditivity is small as we see in the above examples. In Chapter 6, however, we shall find that it cannot be totally ignored.

4.5 Symmetric uniaxial molecules

The treatments up to the previous section are applicable not only to rare-gas atoms but also to the cases in which the polarizability becomes a scalar function $\alpha(\omega)$such as CH_4 and CF_4. On the other hand, in the cases of H_2, N_2, O_2, and CO_2, the parallel and perpendicular components, $\alpha_{\|}(\omega)$, and $\alpha_{\perp}(\omega)$, of the tensor are different; the potential of the dispersion force between two molecules thus depends on the directions of the molecular axes.

Denoting the operators representing the electric dipoles of the two molecules by $\mathbf{p}^{(1)}$ and $\mathbf{p}^{(2)}$, we obtain the potential energy between them as

$$J = r^{-3}(p_x^{(1)}p_x^{(2)} + p_y^{(1)}p_y^{(2)} - 2p_z^{(1)}p_z^{(2)}), \tag{4.30}$$

analogous to (4.12). Here r denotes the intermolecular distance, and the z axis has been chosen along the line connecting the two molecules.

We express the x, y, and z components of $\mathbf{p}$ in terms of p_1, p_2, and p_3 defined relative to the molecular axis. We thus take p_3 to be the component parallel to the molecular axis, and p_2 the component perpendicular to both the molecular axis and the z axis. Let the polar coordinates expressing the directions of molecule 1 and molecule 2 with respect to the z axis be θ_1, φ_1 and θ_2, φ_2 (as in Figure 8.2); we then have

$$\begin{aligned}
p_x^{(1)} &= p_1^{(1)} \cos\theta_1 \cos\varphi_1 - p_2^{(1)} \sin\varphi_1 + p_3^{(1)} \sin\theta_1 \cos\varphi_1 \\
p_y^{(1)} &= p_1^{(1)} \cos\theta_1 \sin\varphi_1 + p_2^{(1)} \cos\varphi_1 + p_3^{(1)} \sin\theta_1 \sin\varphi_1 \\
p_z^{(1)} &= -p_1^{(1)} \sin\theta_1 \qquad\qquad\qquad\qquad\quad + p_3^{(1)} \cos\theta.
\end{aligned}$$

Similar formulas apply for $\mathbf{p}^{(2)}$.

The J expressing the interaction is

$$J = r^{-3} \sum_{j=1}^{3} \sum_{k=1}^{3} p_j^{(1)} T_{jk} p_k^{(2)} \tag{4.31}$$

where

$$T_{jk} = \begin{bmatrix} \cos\theta_1 \cos\varphi_1 & \cos\theta_1 \sin\varphi_1 & -\sin\theta_1 \\ -\sin\varphi_1 & \cos\varphi_1 & 0 \\ \sin\theta_1 \cos\varphi_1 & \sin\theta_1 \sin\varphi_1 & \cos\theta_1 \end{bmatrix} \begin{bmatrix} 1 & 0 & 0 \\ 0 & 1 & 0 \\ 0 & 0 & -2 \end{bmatrix}$$

$$\times \begin{bmatrix} \cos\theta_2 \cos\varphi_2 & -\sin\varphi_2 & \sin\theta_2 \cos\varphi_2 \\ \cos\theta_2 \sin\varphi_2 & \cos\varphi_2 & \sin\theta_2 \sin\varphi_2 \\ -\sin\theta_2 & 0 & \cos\theta_2 \end{bmatrix}.$$

As in Section 4.3, the potential of the dispersion force is

$$W_2(r, \theta_1, \theta_2, \varphi_1 - \varphi_2) = -\mu r^{-6},$$

$$\mu = \sum_j \sum_k (T_{jk})^2 \frac{\hbar}{2\pi} \int_0^\infty \alpha_j^{(1)}(i\omega)\alpha_k^{(2)}(i\omega)d\omega \tag{4.32}$$

where

$$\alpha_j^{(1)}(i\omega) = \frac{2}{\hbar} \sum_{\rho \neq 0} \frac{|\langle 0 | p_j^{(1)} | \rho \rangle|^2 \omega_\rho}{\omega_\rho^2 + \omega^2};$$

$\alpha_j^{(2)}(i\omega)$ may be expressed similarly. Specifically, writing

$$\alpha_1(i\omega) = \alpha_2(i\omega) = \alpha_\parallel(i\omega), \quad \alpha_3(i\omega) = \alpha_\perp(i\omega),$$

and carrying out somewhat complicated calculations, we obtain the following formula for μ:

$$\mu = \frac{\hbar}{3\pi} \int_0^\infty [2\alpha_\perp^{(1)}(i\omega) + \alpha_\parallel^{(1)}(i\omega)]\,[2\alpha_\perp^{(2)}(i\omega) + \alpha_\parallel^{(2)}(i\omega)]\,d\omega$$

$$+ (3\cos^2\theta_1 - 1)\frac{\hbar}{6\pi} \int_0^\infty [\alpha_\parallel^{(1)}(i\omega) - \alpha_\perp^{(1)}(i\omega)]\,[2\alpha_\perp^{(2)}(i\omega) + \alpha_\parallel^{(2)}(i\omega)]\,d\omega$$

$$+ (3\cos^2\theta_2 - 1)\frac{\hbar}{6\pi} \int_0^\infty [2\alpha_\perp^{(1)}(i\omega) + \alpha_\parallel^{(1)}(i\omega)]\,[\alpha_\parallel^{(2)}(i\omega) - \alpha_\perp^{(2)}(i\omega)]\,d\omega$$

$$+ [(\sin\theta_1 \sin\theta_2 \cos(\varphi_1 - \varphi_2) - 2\cos\theta_1 \cos\theta_2)^2 - \cos^2\theta_1 - \cos^2\theta_2]$$

$$\times \frac{\hbar}{2\pi} \int_0^\infty [\alpha_\parallel^{(1)}(i\omega) - \alpha_\perp^{(1)}(i\omega)]\,[\alpha_\parallel^{(2)}(i\omega) - \alpha_\perp^{(2)}(i\omega)]\,d\omega. \tag{4.33}$$

This result expresses μ as a summation of four terms. The second and fourth terms vanish after averaging with respect to directions of atom 1; only the first term remains after averaging with respect to directions of both atom 1 and atom 2.

The result calculated for H_2 molecules by Victor and Dalgarno (G. A. Victor and

A. Dalgarno (1970), *J. Chem. Phys.*, **53**, 1316) is, in atomic units,

$$\mu = 12.38 + 0.62(3\cos^2\theta_1 + 3\cos^2\theta_2 - 2)$$
$$+ 0.20[(\sin\theta_1 \sin\theta_2 \cos(\varphi_1 - \varphi_2) - 2\cos\theta_1 \cos\theta_2)^2$$
$$- \cos^2\theta_1 - \cos^2\theta_2].$$

As in (4.20) we may approximate

$$\alpha_\parallel(i\omega) = \frac{\tilde{\omega}_\parallel^2 \alpha_\parallel}{\tilde{\omega}_\parallel^2 + \omega^2}, \qquad \alpha_\perp(i\omega) = \frac{\tilde{\omega}_\perp^2 \alpha_\perp}{\tilde{\omega}_\perp^2 + \omega^2}$$

in terms of the characteristic frequencies $\tilde{\omega}_\parallel$ and $\tilde{\omega}_\perp$; with the aid of the electrostatic polarizabilities $\alpha_\parallel = 6.30$ and $\alpha_\perp = 4.84$, we then have

$$\hbar\tilde{\omega}_\parallel = 0.60, \quad \hbar\tilde{\omega}_\perp = 0.57 \text{ (in atomic units).}$$

These values are close to the ionization energy, 0.567, of the H_2 molecule.

Generally the characteristic frequencies, $\tilde{\omega}_\parallel$ and $\tilde{\omega}_\perp$, may be taken to be close to each other. The value of $\hbar\tilde{\omega}$, however, is not always so close to the ionization energy. The results of this section and in particular (4.33) will be used in Section 8.3.

Chapter 5

Equation of State for Gases

5.1 Virial coefficients and cluster coefficients

An equation of state for a single-component gas describing the relationship between the pressure P, the temperature T, and the number density n (i.e. the number of molecules in a unit volume) may be expressed as

$$\frac{P}{kT} = n + B(T)n^2 + C(T)n^3 + D(T)n^4 + \ldots, \tag{5.1}$$

where k is the Boltzmann constant; $B(T)$, $C(T)$, and $D(T)$ are called the second, the third, and the fourth virial coefficients. The results of experimental measurements are also summarized usually in this form.

We have noted in Section 1.3 that the second virial coefficient is positive at high temperatures and negative at low temperatures; the temperature T_B, called the Boyle temperature, exists so that for $T > T_B, B(T) > 0$, and for $T < T_B$ $B(T) < 0$. As a quantity related to T_B, we may define

$$b \equiv T_B \left(\frac{dB(T)}{dT}\right)_{T=T_B} = \left(\frac{dB}{d \ln T}\right)_{T=T_B}. \tag{5.2}$$

The magnitude of b is of the order of the molecular volume. (For a gas obeying the van der Waals equation of state, the b in (1.1) agrees with this quantity; hence the same symbol is used.) With the aid of this b, we may introduce dimensionless quantities,

$$B^* \equiv B/b, \quad c \equiv C/b^2, \quad D^* \equiv D/b^3, \ldots .$$

These dimensionless quantities may be regarded as mutually of the same order of magnitude.

In Table 5.1 we list the values of T_B and b, together with the values of the critical temperatures T_c and the ratios T_c/T_B. As the number of atoms in a molecule (the hydrogen atoms are usually discounted, however) increases the ratio T_c/T_B increases; this will be explained later (Section 5.7).

As a gas is compressed at a temperature lower than the critical temperature T_c, it will be liquefied. During the process of such liquefaction there will be stages in which combinations of two or three molecules are mixed with single molecules. Such a gas may thus be regarded as an ideal-gas mixture of single molecules,

Table 5.1 Boyle temperature T_B, volume b, and critical temperature T_c

	T_B(K)	b(Å^3)	T_c(K)	T_c/T_B
Ne	122	35.4	44.5	0.36
Ar	407	67.5	150.7	0.37
Kr	570	84.6	209.4	0.37
Xe	774	118	289.8	0.37
CH_4	506	91.3	190.6	0.38
O_2	405	66.9	154.7	0.38
N_2	323	89.1	126.2	0.39
CO_2	700	122	304.2	0.43
CF_4	517	174	227.6	0.44
$C(CH_3)_4$	951	425	433.8	0.46

two-molecule clusters, three-molecule clusters, etc. Writing the number density of single molecules as z, and those of l-molecule clusters as $b_l z^l$ ($l = 2, 3, \ldots$), we have the expression

$$\frac{P}{kT} = \sum_{l=1}^{\infty} b_l z^l, \quad n = \sum_{l=1}^{\infty} l b_l z^l \quad (b_1 \equiv 1). \tag{5.3}$$

Here the $b_l = b_l(T)$ are functions of temperature; these are called the l-molecule *cluster coefficients.*

At low temperatures, the physical picture of (5.3) is thus clear; adopting its form over the entire temperature domain would be equivalent to the virial expansion (5.1) as an equation of state for the gas. It is nothing but a kind of a parametric representation with z as the parameter. Elimination of z from the two equations yields the virial expansion,

$$P/kT = n - b_2 n^2 + (4b_2^2 - 2b_3)n^3 + \ldots .$$

The relations expressing the cluster coefficients b_l in terms of the virial coefficients are:

$$\begin{aligned} b_2 &= -B, \quad b_3 = 2B^2 - \tfrac{1}{2}C, \\ b_4 &= -\tfrac{16}{3}B^3 + 3BC - \tfrac{1}{3}D, \\ b_5 &= \tfrac{50}{3}B^4 - 15B^2C + \tfrac{9}{8}C^2 + \tfrac{8}{3}BD - \tfrac{1}{4}E \ldots . \end{aligned} \tag{5.4}$$

For the cluster coefficients, we also define dimensionless quantities,

$$b_l^* \equiv b_l / b^{l-1}, \tag{5.5}$$

in terms of the b given by (5.2).

A remarkable feature may be noted in the numerical coefficients on the right-hand side of various formulas in (5.4). The contribution of the fourth virial coefficient D in b_4 is smaller by an order of magnitude than those of B and C. A

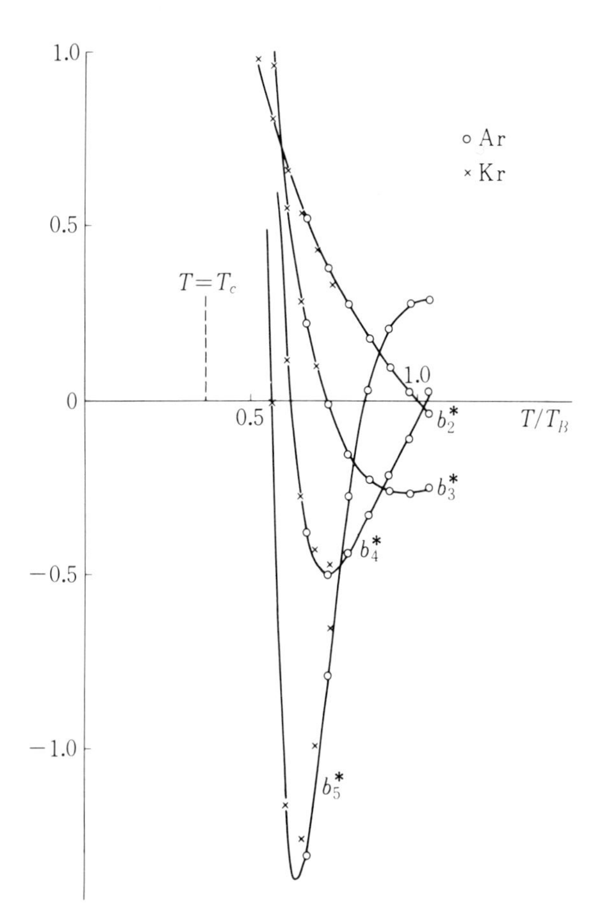

Figure 5.1 Dimensionless cluster coefficients obtained from the virial coefficients of argon and krypton

similar feature persists in b_5; the contribution of the fifth virial coefficient E is further smaller by an order of magnitude. In fact, the accuracy of the 'measured values' of those virial coefficients obtained from experimental data decreases by nearly an order of magnitude in the order of B, C, D, E. For the reasons stated above, however, the accuracy of the cluster coefficients would not decrease substantially; thus it is still meaningful to treat b_l up to those large l (see Section 5.3).

The curves in Figure 5.1 depict the relation between $b_l{}^*$ and T/T_B for argon and krypton. As we see here the values for different gases are plotted on a common curve when expressed in terms of the reduced dimensionless quantities; this is a manifestation of the law of corresponding states.

For the polyatomic molecules, the law of corresponding states does not hold with a good accuracy even if only nonpolar molecules are considered. As an

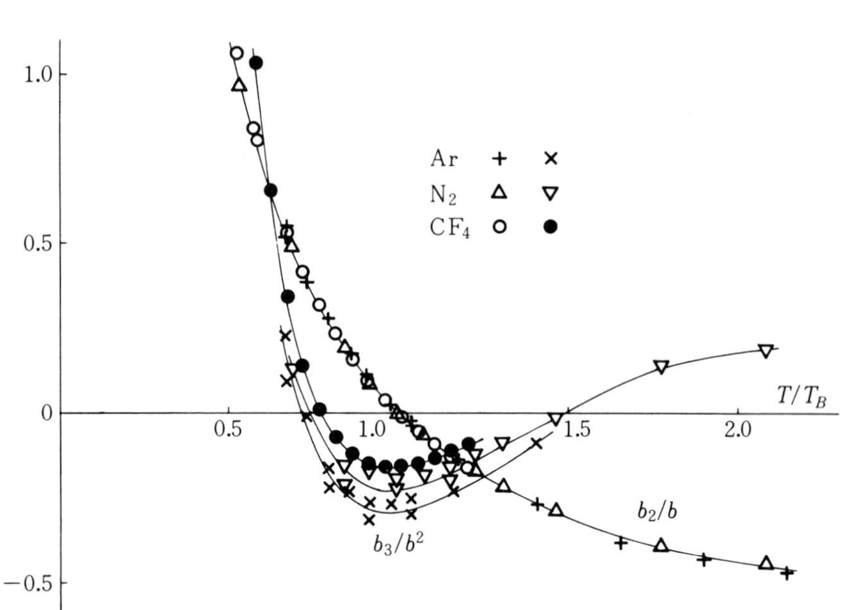

Figure 5.2 Examples of measured values of the two-molecule and the three-molecule cluster coefficients: +, △, ○, two-molecule; x, ▽, ● three-molecule

example, we compare in Figure 5.2 the cases of argon, nitrogen, and carbon tetrafluoride.

5.2 The Gamma function

In this book the following formula is frequently used:

$$\int_{-\infty}^{\infty} \exp(-x^2)dx = \sqrt{\pi}. \tag{5.6}$$

To derive this, we multiply the left-hand side by the same expression and calculate

$$\int_{-\infty}^{\infty}\int_{-\infty}^{\infty} \exp[-(x^2+y^2)]dxdy = \int_0^{\infty} \exp(-r^2)2\pi r dr = \pi.$$

Here we have regarded the integration variables x and y as ordinary x and y coordinates and transformed them into polar coordinates. Equation (5.1) may also be written as

$$\int_0^{\infty} e^{-t}t^{-1/2}dt = \sqrt{\pi}. \tag{5.7}$$

In general the gamma function $\Gamma(x)$ is defined as

$$\Gamma(x) = \int_0^{\infty} e^{-t}t^{x-1}dt \quad x > 0. \tag{5.8}$$

As we see with the aid of partial integrations, a recursive formula

$$\Gamma(x+1) = x\Gamma(x) \tag{5.9}$$

holds. In particular, when x is an integer n

$$\Gamma(n+1) = n! \quad \Gamma(1) = 1.$$

For a half-integer, we have, with the aid of (5.7),

$$\Gamma(1/2) = \sqrt{\pi}, \quad \Gamma(3/2) = (\sqrt{\pi})/2, \quad \Gamma(5/2) = 3(\sqrt{\pi})/4, \quad \ldots .$$

The gamma function with a complex variable is defined through analytic continuation of (5.8). For such a function, (5.9) likewise holds as long as $x \neq 0$, $-1, -2, \ldots$. Consequently, for a real number x in the domain $-1 < x < 0$, we have

$$\Gamma(x) = x^{-1}\Gamma(x+1).$$

For example $\Gamma(-1/2) = -2\sqrt{\pi}$.

Using the integral representation (5.8), we may derive an asymptotic form (Stirling's formula) of $n!$,

$$n! \sim \sqrt{(2\pi n)}n^n e^{-n}, \tag{5.10}$$

for a large integer n. First by setting

$$\int_0^\infty e^{-t}t^n dt = \int_0^\infty e^{f(t)}dt,$$

we have

$$f(t) = -t + n \ln t.$$

This $f(t)$ has a pronounced maximum at $t = n$ and $f''(n) = -1/n$. Hence the major contribution to the integration comes from the domain in which an approximation

$$f(t) \doteqdot f(n) - (t-n)^2/2n$$

holds. We thus obtain

$$n! \sim e^{-n}n^n \int_{-\infty}^{\infty} \exp\left(-\frac{(t-n)^2}{2n}\right) d(t-n) = \sqrt{(2\pi n)}n^n e^{-n}.$$

In the integration above, we have used the fact that $f(t)$, as a function of a real variable t, has a maximum at $t = n$. On the complex t plane, $t = n$ is a saddle point of $|e^{f(t)}|$; the foregoing path of integration corresponds to that passing through the saddle point where the integrand decreases most steeply. The method of calculating an asymptotic form of an integration by selecting the integration path passing through the saddle point where the integrand decreases most steeply, namely the path of steepest descent, is called the *method of steepest descent* or the saddle-point method.

Of course, the path of steepest descent does not always correspond to the real axis; nor does it always form a straight line. In the following section we will find such an example.

5.3 Zeros of the cluster coefficients and the critical temperature

Figure 5.1 shows the temperature dependence of the cluster coefficients b_l. Let us now consider the temperatures T at which $b_l(T) = 0$, i.e. the zeros of b_l. As l increases, the minimum value of the zeros for b_l decreases and accumulates to a point. Denoting the temperature at such an accumulation point by T_a, we find that all the b_l are positive for temperatures equal to or smaller than T_a.

It is possible to prove thermodynamically that such an accumulation point T_a is equal to the critical temperature T_c (T. Kihara and J. Okutani (1971), *Chem. Phys. Letters*, **8**, 63).

For such a proof it may be sufficient to show the following two properties: (i) for a large l, $b_l(T_c) > 0$; (ii) for a T greater than T_c there exists an infinite number of zeros of the cluster coefficients between T_c and T. It then follows that, if $T_a < T_c$, (i) would not be true; if $T_c < T_a$, (ii) would not be true.

From (5.3) or

$$n = \sum_{l=1}^{\infty} l b_l z^l, \qquad \frac{P}{kT} = \sum_{l=1}^{\infty} b_l z^l, \tag{5.11}$$

we have at a given temperature T

$$l b_l(T) = \frac{1}{2\pi \mathrm{i}} \oint \frac{n}{z^{l+1}} dz, \tag{5.12}$$

where the integration path is to encircle the origin in the positive direction on the complex z plane.

The parameter z may now be expressed as

$$z = \mathrm{e}^{K(T,\, n)} \quad \text{i.e.} \quad \ln z = K(T, n). \tag{5.13}$$

Hence at a constant temperature,

$$z^{-1} dz = (\partial K/\partial n) dn. \tag{5.14}$$

For the coefficient $\partial K/\partial n$, combining (5.14) and the formula

$$\frac{1}{kT}\left(\frac{\partial P}{\partial n}\right)_T = \frac{n}{z}\frac{\partial z}{\partial n}$$

obtainable from (5.11), we have the relation

$$\frac{\partial K}{\partial n} = \frac{1}{kT}\frac{1}{n}\left(\frac{\partial P}{\partial n}\right)_T. \tag{5.15}$$

Now, elimination of z from (5.12) with the aid of (5.13) and (5.14) yields

$$l b_l(T) = \frac{1}{2\pi \mathrm{i}} \oint n \mathrm{e}^{-lK(T,\, n)} \frac{\partial K}{\partial n} dn = \frac{-1}{2\pi \mathrm{i}}\frac{1}{l} \oint n \frac{\partial}{\partial n} \mathrm{e}^{-lK(T,\, n)} dn,$$

or

$$l^2 b_l(T) = \frac{1}{2\pi \mathrm{i}} \oint \mathrm{e}^{-lK(T,\, n)} dn \tag{5.16}$$

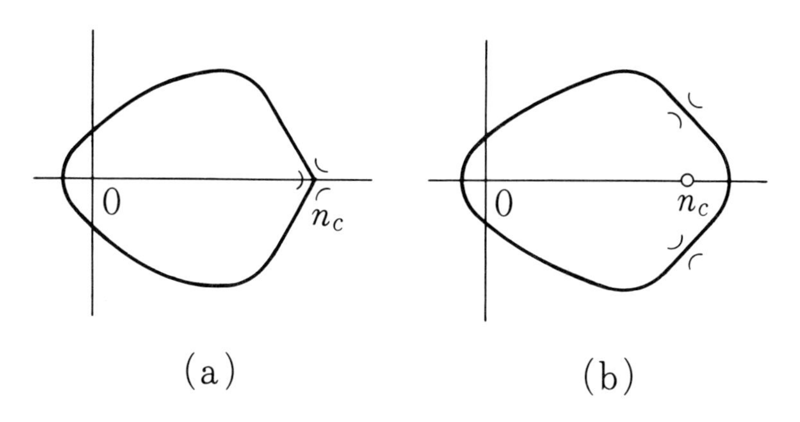

Figure 5.3

where the path of integration is to encircle the origin in the positive direction on the complex n plane. (Near the origin, n and z coincide with each other.) The asymptotic form of b_l for large l may be obtained by deforming the integration path of (5.16) in such a way as to make the path of steepest descent of the integrand. The position of the saddle point is given by $\partial K/\partial n = 0$, that is, $(\partial P/\partial n)_T = 0$.

At the critical temperature $T = T_c$, n_c corresponding to the critical point is the position of the saddle point. In its vicinity,

$$K(T_c, n) = K(T_c, n_c) - \alpha(n_c - n)^3, \tag{5.17}$$

where α is a positive constant on account of (5.15). Choosing the path of integration as shown in Figure 5.3a, we obtain for $l \gg 1$

$$\begin{aligned} l^2 b_l(T_c) &= \exp[-lK(T_c, n_c)] \frac{1}{2\pi i} (e^{2\pi i/3} - e^{-2\pi i/3}) \int_0^\infty \exp[-l\alpha\xi^3] d\xi \\ &= \frac{\sqrt{3}}{2\pi} \Gamma\left(\frac{4}{3}\right) \frac{1}{(l\alpha)^{1/3}} \exp[-lK(T_c, n_c)]. \end{aligned} \tag{5.18}$$

Consequently we have proved that $b_l(T_c)$ is positive for $l \gg 1$.

In reality, the thermodynamic relationship between T, n, and P is not always analytic in the vicinity of the critical point as (5.17) would imply. Along the isotherm at $T = T_c$, in particular,

$$|P - P_c| = O(|n_c - n|^\delta).$$

holds. The magnitude of the power index δ is 4.4 ± 0.4 for xenon. Equation (5.17) is then replaced by

$$K(T_c, n) = K(T_c, n_c) - \alpha(n_c - n)^\delta,$$

and (5.18) becomes

$$l^2 b_l(T) = \frac{\sin(\pi/\delta)\Gamma(1 + 1/\delta)}{\pi(l\alpha)^{1/\delta}} \exp[-lK(T_c, n_c)],$$

Since no essential change would be involved in so doing, we shall adopt a simplest power index in the following.

When T is slightly greater than T_c, the leading terms of the expansion are

$$K(T, n) = K(T, n_c) - \alpha(n_c - n)^3 - \gamma(T - T_c)(n_c - n),$$

where α and γ are positive constants. The position n_0 of the saddle point is determined from $\partial K/\partial n = 0$, i.e.

$$3\alpha(n_c - n)^2 + \gamma(T - T_c) = 0.$$

Its solution is

$$n_0 = n_c \pm i\rho, \quad \rho = [(T - T_c)\gamma/3\alpha]^{1/2}. \tag{5.19}$$

Carrying out a similar calculation for the integration path of Figure 5.3b we obtain

$$l^2 b_l(T) = \left(\frac{1}{3\pi l\alpha\rho}\right)^{1/2} \sin\left(\frac{3\pi}{4} - 2l\alpha\rho^3\right) \exp[-lK(T, n_c)]. \tag{5.20}$$

where

$$l\alpha\rho^3 \gg 1.$$

From this expression we find that an infinite number of zeros exist between T_c and T.

The limit of convergence for the two power series in (5.11) is the following. For temperatures equal to or smaller than the critical temperature T_c, both are power series of positive terms; hence the limit of convergence is the singularity of the functions represented by the power series. This singularity is the critical point at $T = T_c$.

For temperatures smaller than T_c, it may be clear that the onset of liquefaction corresponds to such a singular point. If it is in a state of supercooling, we may retain an appropriate number of terms and cut off those remaining; such a power series may then approximately describe such a metastable state. In this sense the cluster expansion is an asymptotic series.

For temperatures greater than T_c, the limit of convergence for the two power series does not necessarily represent the singular point of the equation of state. In fact, as may be clear in (5.20) at temperatures slightly greater than T_c the density corresponding to the limit of convergence differs only little from the density n_c at the critical point.

Combining the theorem that the critical temperature is determined from the distribution of zeros of the cluster coefficients in low temperature domain and the property that the four-molecule, five-molecule, ... cluster coefficients do not substantially depend on the fourth, fifth, ... virial coefficients, we may be able to calculate approximately the value of the critical temperature from only the second and third virial coefficients, $B(T)$ and $C(T)$. Thus, neglecting the terms beginning with $D(T)n^4$ in (5.1) we set

$$\frac{P}{kT} = n + B(T)n^2 + C(T)n^3,$$

whence we may calculate an approximate value T_c' of the critical temperature by eliminating n from the relations,

$$\frac{d}{dn}\left(\frac{P}{kT}\right) = 0, \quad \frac{d^2}{dn^2}\left(\frac{P}{kT}\right) = 0.$$

The result is

$$3C(T_c') = [B(T_c')]^2. \tag{5.21}$$

The errors involved in this estimation are of the order of 10% (see Figure 5.6).

5.4 The canonical distribution

Suppose a single-component gas to be contained in a volume V; the number of molecules is N, and we assume that it is in thermodynamic equilibrium with a thermal reservoir at an absolute temperature T. A macroscopic state of such a system corresponds to a kind of average over extremely numerous microscopic states. The canonical distribution describes the probability distribution of such numerous microscopic states.

From the quantum-mechanical point of view, such a microscopic state may be regarded as a quantum state with an energy eigenvalue. Setting the probability of a quantum-mechanical state belonging to an energy eigenvalue E_n as w_n, we write the canonical distribution as

$$w_n = w(E_n) = A \exp(-E_n/kT), \tag{5.22}$$

where k is the Boltzmann constant. The factor A is determined from the condition,

$$\sum_n w_n = 1.$$

From the point of view of the classical theory we may argue in the following way. Assuming that the gas molecules do not have internal degrees of freedom, we may express the energy $E(p, q)$ as a function of p and q, where q collectively describes the positions of N molecules and p describes the momenta. Denoting by $\rho(p, q)dpdq$ the probability that microscopic states are found in the volume element $dpdq$ in the $6N$-dimensional phase space of p and q, we may write canonical distribution as

$$\rho(p,q) = A\mathrm{e}^{-E(p,q)/kT}, \tag{5.23}$$

where the factor A is again determined from

$$\rho(p,q)dpdq = 1.$$

Equations (5.22) and (5.23) are derived in the following way. Suppose that a system consisting of two parts, (1) and (2), is in thermodynamic equilibrium with a thermal reservoir. Let $w^{(1)}(E_n')$ be the probability that (1) is in the quantum state with energy E_n', and $w^{(2)}(E_n'')$, the probability that (2) is in the quantum state with energy E_n''. Since the probability $w(E_n)$ that the combined system of (1) and (2) is in a quantum state with energy

$$E_n = E_n' + E_n''$$

is given by the product of those probabilities,

$$w(E_n) = w^{(1)}(E_n')w^{(2)}(E_n''),$$

we have

$$\ln w(E_n) = \ln w^{(1)}(E_n') + \ln w^{(2)}(E_n'').$$

This relation is satisfied if and only if

$$\ln w^{(1)}(E_n') = \alpha^{(1)} - \beta E_n', \quad \ln w^{(2)}(E_n'') = \alpha^{(2)} - \beta E_n'',$$
$$\ln w(E_n) = \alpha - \beta E_n, \quad \alpha = \alpha^{(1)} + \alpha^{(2)}.$$

We thus find that $\ln w(E)$ is a linear function of E, i.e.

$$w(E_n) = A \exp(-\beta E_n).$$

The β introduced here must be common to all the systems that are in thermodynamic equilibrium with the same thermal reservoir. Similarly we have

$$\rho(p,q) = A e^{-\beta E(p,q)}.$$

To show that $\beta = 1/kT$, we may recall the pressure of an ideal gas, for example. Let m be the mass of a molecule; the x axis is chosen in the direction perpendicular to the wall of the container. The pressure P is equal to the average of a product between the change of the momentum $2p_x$ of a molecule when it collides with the wall and the number of such collisions $(N/2V)p_x/m$ per unit time and per unit area:

$$\begin{aligned} P &= \frac{N}{V} \int_0^\infty \frac{p_x^2}{m} \exp(-\beta p_x^2/2m)dp_x \Big/ \int_0^\infty \exp(-\beta p_x^2/2m)dp_x \\ &= -2\frac{N}{V}\frac{\partial}{\partial\beta} \ln \int_0^\infty \exp(-\beta p_x^2/2m)dp_x \\ &= \frac{N}{V}\frac{1}{\beta}. \end{aligned}$$

Comparing this with $PV = NkT$, we obtain $\beta = 1/kT$.

According to the canonical distribution (5.22), the average value $\bar{E}$ of the energy E_n is given by

$$\bar{E} = \sum_n w_n E_n = \sum_n E_n w(E_n).$$

This average value $\bar{E}$ is the internal energy of the system. The instantaneous values of the energy of the system fluctuate slightly above and below such an average value. Let us look at the number of quantum-states contained in such a range of fluctuations. The number $\Delta\Gamma$ of these quantum states is given by

$$w(\bar{E})\Delta\Gamma = 1. \tag{5.24}$$

When a macroscopic state corresponds to $\Delta\Gamma$ quantum states, the entropy S of the system is

$$S = k \ln \Delta\Gamma. \tag{5.25}$$

From (5.24),

$$S = -k \ln w(\bar{E}),$$

or since $\ln w(\bar{E})$ is equal to the average of $\ln w(\bar{E}_n)$,

$$S = -k \sum_n w_n \ln w_n.$$

Substitution of (5.22) into w_n on the right-hand side yields

$$S = -k \ln A + \bar{E}/T,$$

or with the aid of the free energy F,

$$\ln A = F/kT, \quad F = \bar{E} - TS.$$

Equation (5.22) is thus expressed as

$$w_n = \exp[(F - E_n)/kT]. \tag{5.26}$$

From the condition $\Sigma w_n = 1$,

$$F = -kT \ln \sum_n \exp(-E_n/kT). \tag{5.27}$$

Consequently, once the state sum or the partition function $\Sigma \exp(-E_n/kT)$ is obtained, the free energy F can be calculated as a function of T, V, and N with the aid of (5.27); the pressure P, the chemical potential μ, and so on, can then be found through the thermodynamic relationship,

$$dF = -SdT - PdV + \mu dN. \tag{5.28}$$

When the gas molecules have no internal degrees of freedom, the free energy corresponding to (5.23) is given by

$$F = -kT \ln\left[\frac{1}{N!}\int e^{-E(p,q)/kT}\frac{dpdq}{h^{3N}}\right]. \tag{5.29}$$

The reason that the infinitesimal volume in the $6N$-dimensional phase space is divided by the $3N$th power of the Planck constant h may be explained in the following way. We first recall Bohr's condition of quantization,

$$\oint pdq = (n + \tfrac{1}{2})h, \quad n = 0, 1, 2, \ldots,$$

for a one-dimensional periodic motion of a particle. The number of quantum states corresponding to (or contained in) a surface element in the two-dimensional phase space under these circumstances is given by the area of the surface element divided by h. We may similarly treat the cases in which N molecules perform three-dimensional motion. Next, the reason for dividing the integral in (5.29) by $N!$ is the following. Since we are considering a single-component gas, N particles are equivalent. Hence, the state in which the first particle is at the position A in the phase space and the second particle at the position B represents the same quantum state as that in which the positions of the two particles are interchanged. Since the integration by itself amounts to treating those as if they were different states, we must divide it by $N!$.

As a special example, let us calculate (5.29) for an ideal gas. Denoting the mass of a particle by m, we have

$$e^{-F/kT} = \frac{V^N}{N!}\left[\iiint_{-\infty}^{\infty} \exp\{-(p_x^2 + p_y^2 + p_z^2)/2mkT\} \frac{dp_x dp_y dp_z}{h^3}\right]^N .$$

With the aid of the formula (5.6),

$$\int_{-\infty}^{\infty} \exp(-x^2)dx = \sqrt{\pi},$$

we then find

$$F = -kT \ln\left[\frac{V^N}{N!}\left(\frac{mkT}{2\pi\hbar^2}\right)^{3N/2}\right].$$

Since $N \gg 1$, we may use Stirling's formula to replace $\ln N!$ by $N \ln N - N$:

$$F = -NkT \ln\left[\frac{eV}{N}\left(\frac{mkT}{2\pi\hbar^2}\right)^{3/2}\right],$$

where $\ln e = 1$. From (5.28), the pressure is

$$P = -\left(\frac{\partial F}{\partial V}\right)_{T,N} = \frac{NkT}{V}.$$

Naturally the equation of state for the ideal gas results. The chemical potential μ is

$$\mu = \left(\frac{\partial F}{\partial N}\right)_{T,V} = -kT \ln\left[\frac{V}{N}\left(\frac{mkT}{2\pi\hbar^2}\right)^{3/2}\right],$$

or

$$\frac{N}{V} = \left(\frac{mkT}{2\pi\hbar^2}\right)^{3/2} e^{\mu/kT}. \tag{5.30}$$

The right-hand side of this formula is a kind of quantity called the *activity*, which may be used in lieu of the chemical potential. For the ideal gas the activity as such is a quantity easy to understand intuitively.

5.5 The grand canonical distribution

The word 'canonical' in the canonical distribution may mean 'free of restriction'. In fact, by maintaining equilibrium with a thermal reservoir at temperature T, restriction on the energy of the system has been removed.

On the other hand, the number of molecules has been fixed at N. The distribution in which the condition of a constant number of molecules as well as that of a constant energy has been removed is called the grand canonical distribution.

Considering a single-component gas in a volume V, we assume that the system is in thermodynamic equilibrium with environment characterized by the temperature T and the chemical potential μ. This environment acts both as a thermal reservoir and as a source of particles, exchanging particles with the system (for example, through a small pore). A macroscopic state of the system is an average of extremely numerous microscopic states. A microscopic state may be regarded as a quantum state belonging to an energy eigenvalue E_{nN} with number of particles, N. Let w_{nN} be the probability of such a quantum state, that is,

$$w_{nN} = w(E_{nN}, N), \tag{5.31}$$

with the condition

$$\sum_N \sum_n w_{nN} = 1. \tag{5.32}$$

The thermodynamic variable N appearing in (5.28) is the average value $\bar{N}$ of N in the above formula. Introducing a new thermodynamic function Ω defined by

$$\Omega = F - \mu\bar{N}, \tag{5.33}$$

we have from (5.28)

$$d\Omega = -SNT - PdV - \bar{N}d\mu. \tag{5.34}$$

We calculate this Ω as a function of T, μ, and V.

We expect that the grand canonical distribution corresponds to the canonical distribution (5.26) in which F is replaced by $\Omega + \mu N$:

$$w_{nN} = \exp[(\Omega + \mu N - E_{nN})/kT]. \tag{5.35}$$

In fact, the entropy S in this distribution is

$$S = -k \ln w(\bar{E}, \bar{N}) = (-\Omega - \mu\bar{N} + \bar{E})/T,$$

whence

$$\Omega = \bar{E} - TS - \mu\bar{N} = F - \mu\bar{N}.$$

Equation (5.33) thus obtains.

Applying the condition (5.32) to the probability distribution (5.35), we have

$$\Omega = -kT \ln \left[\sum_N e^{\mu N/kT} \sum_n \exp(-E_{nN}/kT) \right]. \tag{5.36}$$

When the gas molecules have no internal degrees of freedom, similarly to the previous section, we obtain

$$\Omega = -kT \ln \left[\sum_N e^{\mu N/kT} \frac{1}{N!} \int \exp\{-E_N(p, q)/kT\} \frac{dpdq}{h^{3N}} \right], \tag{5.37}$$

where $E_N(p, q)$ represents the energy at the number of particles, N.

In particular for an ideal gas with the mass of a particle m, the integral in the foregoing formula becomes $V^N\ (mkT/2\pi\hbar^2)^{3N/2}$. Hence,

$$\Omega = -kT \left(\frac{mkT}{2\pi\hbar^2}\right)^{3/2} e^{\mu/kT} V.$$

From (5.34)

$$\frac{\bar{N}}{V} = -\frac{1}{V}\frac{\partial\Omega}{\partial\mu} = \left(\frac{mkT}{2\pi\hbar^2}\right)^{3/2} e^{\mu/kT}. \tag{5.38}$$

This corresponds to (5.30). Furthermore,

$$P = -\frac{\partial\Omega}{\partial V} = kT \left(\frac{mkT}{2\pi\hbar^2}\right)^{3/2} e^{\mu/kT}. \tag{5.39}$$

Elimination of the activity from those two formulas yields the equation of state for an ideal gas.

Problem

The entropy S is defined by the form of (5.25), i.e.

$$S = k \ln \Delta\Gamma,$$

where $\Delta\Gamma$ is the number of microscopic states corresponding to a macroscopic state. In the grand canonical distribution, the number N of molecules distributes in the range of fluctuations ΔN; the $\Delta\Gamma$ then is given by the product between $\Delta\Gamma$ in the canonical distribution and ΔN. Explain the reason why the same value of entropy is obtained for either case of the distributions irrespective of that difference.

Solution

Since ΔN is of the order of $\sqrt{N}$, the difference in $\ln \Delta\Gamma$ is of the order of $\ln N$. This is negligible compared with N.

Incidentally, the formula $\ln N! = N \ln N - N$ used for the canonical distribution neglects just this difference in Stirling's formula (Section 5.2),

$$\ln N! = N \ln N - N + \tfrac{1}{2} \ln N + O(1).$$

The consideration above is therefore consistent.

5.6 Cluster integrals

On the basis of the grand canonical distribution described in the previous section, we calculate the equation of state for a single-component rare gas in a form of a power-series expansion. In (5.37) i.e.

$$e^{-\Omega/kT} = \sum_N \frac{1}{N!} e^{\mu NkT} \int \exp[-E_N(p,q)/kT] \frac{dpdq}{h^{3N}}$$

$E_N(p, q)$ represents the energy of the system composed of N atoms. For $N = 0$, of course, $E_N(p, q) = 0$. For $N = 1$, it is just the kinetic energy of an atom:

$$E_1(p,q) = p^2/2m,$$

where m is the mass of a rare-gas atom. For $N = 2$, it is equal to the kinetic energy of two atoms plus the interaction energy:

$$E_2(p, q) = \sum_{a=1}^{2} \frac{p_a^2}{2m} + U_{12}.$$

For $N = 3$,

$$E_3(p, q) = \sum_{a=1}^{3} \frac{p_a^2}{2m} + U_{123},$$

where U_{123} is the interaction energy between three atoms. U_{123} is approximately given by $U_{12} + U_{13} + U_{23}$; there exists a difference, however, arising from nonadditivity of the molecular potential.

As for the activity considered in the foregoing sections, we define

$$z \equiv e^{\mu/kT} \iiint \exp[-(p_x^2 + p_y^2 + p_z^2)/2mkT] \frac{dp_x dp_y dp_z}{h^3} = \left(\frac{mkT}{2\pi\hbar^2}\right)^{3/2} e^{\mu/kT}.$$

We then have

$$e^{-\Omega/kT} = 1 + zV + \frac{z^2V}{2!} \int \exp(-U_{12}/kT) d\mathbf{r}_2$$

$$+ \frac{z^3V}{3!} \iint \exp(-U_{123}/kT) d\mathbf{r}_2 d\mathbf{r}_3 + \ldots .$$

The range of integration for the position of each atom is over the volume V of the system. Since the interaction energy depends only on the relative positions of the atoms, the integration with respect to atom 1 produces simply a factor V.

The potential U_{12} between two atoms does not vanish only when the two atoms are close to each other. Thus, writing

$$b_2 = \frac{1}{2!} \int [\exp(-U_{12}/kT) - 1] d\mathbf{r}_2, \tag{5.40}$$

we find that $|b_2|$ is a quantity of the order of the atomic volume. Similarly the integrand of

$$b_3 = \frac{1}{3!} \iint [\exp(-U_{123}/kT) - \exp(-U_{12}/kT) - \exp(-U_{13}/kT) - \exp(-U_{23}/kT) + 2]\, d\mathbf{r}_2 d\mathbf{r}_3 \tag{5.41}$$

is nonzero only when the three atoms are close to each other. For, when atom 3 is sufficiently far away from atoms 1 and 2, U_{123} is equal to U_{12}; U_{13} and U_{23} thus vanish. Consequently, $|b_3|$ is a quantity of the order of the square of the atomic volume. In terms of such $b_2, b_3, \ldots$, we write

$$e^{-\Omega/kT} = 1 + zV + \left(\frac{V}{2!} + b_2\right) z^2 V + \left(\frac{V^2}{3!} + b_2 V + b_3\right) z^3 V + \ldots,$$

whence we obtain

$$\Omega = -VkT(z + b_2 z^2 + b_3 z^3 + \ldots) = -VkT \sum_{l=1}^{\infty} b_l z^l \quad (b_1 \equiv 1).$$

In general b_l represents an integration over a configuration of l nearby atoms, namely a cluster configuration; it is thus called the l-molecule *cluster integral* (J. E. Mayer 1937).

Regarding Ω as a function of V, T, and μ, we calculate the pressure as

$$P = -\frac{\partial \Omega}{\partial V} = kT \sum_{l=1}^{\infty} b_l z^l;$$

the number of particles $\bar{N}$ as a thermodynamic variable is given by

$$\bar{N} = -\frac{\partial \Omega}{\partial \mu} = V \sum_{l=1}^{\infty} l b_l z^l.$$

Setting the number density $\bar{N}/V$ of the particles as n, we have

$$\frac{P}{kT} = \sum_{l=1}^{\infty} b_l z^l, \quad n = \sum_{l=1}^{\infty} l b_l z^l. \tag{5.42}$$

We thus obtain the equation of state in the same form as (5.3); it is clear that the parameter z there has the meaning of the activity and that the cluster coefficient b_l is given by the cluster integral.

With the aid of the relation (5.4) between the cluster coefficients and the virial coefficients, the expression for the second virial coefficient $B(T)$ is

$$B(T) = \frac{1}{2} \int [1 - \exp(-U_{12}/kT)]\, d\mathbf{r}_2. \tag{5.43}$$

When the potential is additive, that is,

$$U_{123} = U_{12} + U_{13} + U_{23},$$

the third virial coefficient is expressed as

$$C(T) = \frac{1}{3} \iint [1 - \exp(-U_{12}/kT)]\,[1 - \exp(-U_{13}/kT)] \times [1 - \exp(-U_{23}/kT)]\, d\mathbf{r}_2 d\mathbf{r}_3. \tag{5.44}$$

A proof of this formula may be obtained through a series expansion.

For those temperatures at which kT is smaller than the depth of the potential U_{12}, the negative portion of the potential produces a substantial contribution in (5.43) and thus $B < 0$; at those temperatures for which kT is sufficiently greater than the potential depth, the positive portion of the potential produces a significant contribution and $B > 0$. In other words, the sign of B is determined by the attractive portion of the intermolecular potential at low temperatures and by the repulsive portion at high temperatres.

Although we have thus far treated the rare gases, the same expressions for the virial coefficients and cluster integrals also apply for polyatomic molecules as long as they are approximated by a spherically symmetric shape. For nonspherically symmetric molecules with a fixed shape, we may carry out both integration with respect to the position of a fixed point (e.g. the centre) in the molecule and averaging with respect to the molecular orientation. Instead of (5.43), the second virial coefficient now is given by

$$B(T) = \frac{1}{2} \left\langle \int [1 - \exp(-U_{12}/kT)]\, d\mathbf{r}_2 \right\rangle, \tag{5.45}$$

where $\langle \dots \rangle$ means the average with respect to the orientation of the molecule 2; we may similarly treat $C(T)$.

5.7 Spherically symmetric square-well potential

A simple model of the intermolecular potential for which the second and third virial coefficients as well as the two-molecule and three-molecule cluster integrals may be calculated in terms of elementary functions is offered by a 'spherically symmetric square-well potential'

$$U(r) = \begin{cases} \infty & r < \sigma \\ -\epsilon < 0 & \sigma < r < g\sigma \\ 0 & g\sigma < r \end{cases} \tag{5.46}$$

(Figure 5.4). Here σ is the summation of the radii of the two molecules; for a single component gas it means the molecular diameter.

From (5.43) the second virial coefficient $B(T)$ is

$$B(T) = 2\pi \int_0^\infty (1 - \mathrm{e}^{-U(r)/kT}) r^2\, dr.$$

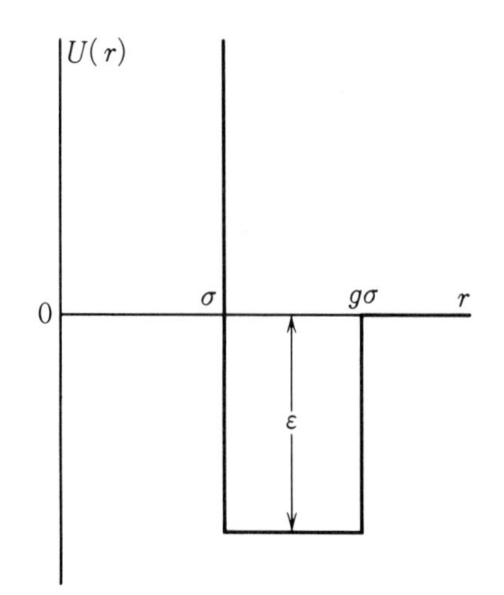

Figure 5.4

On account of (5.46),

$$1 - e^{-U(r)/kT} = \begin{cases} 1 & r < \sigma \\ -x & \sigma < r < g\sigma \\ 0 & g\sigma < r \end{cases}$$

where

$$x = e^{\epsilon/kT} - 1. \tag{5.47}$$

We thus obtain

$$B = 4v_0[1 - (g^3 - 1)x], \quad v_0 \equiv (\pi/6)\sigma^3 \tag{5.48}$$

where v_0 corresponds to the molecular volume.

The Boyle temperature T_B and $b \equiv (dB/d\ln T)_{T=T_B}$ used for normalization of volume are related to the constants in the potential via

$$\frac{\epsilon}{kT_B} = \ln \frac{g^3}{g^3 - 1}, \quad \frac{b}{4v_0} = g^3 \ln \frac{g^3}{g^3 - 1}. \tag{5.49}$$

While the second virial coefficient has been expressed as a linear function of x, the third virial coefficient becomes a cubic function of x; from (5.44)

$$3C(T) = I^{(0)} - 3xI^{(1)} + 3x^2 I^{(2)} - x^3 I^{(3)}.$$

Here $I^{(i)}$ represents integrations $\iint d\mathbf{r}_2 d\mathbf{r}_3$ over the following domains:

$$
\begin{array}{lll}
I^{(0)}: r_{12} < \sigma, & r_{13} < \sigma, & r_{23} < \sigma \\
I^{(1)}: \sigma < r_{12} < g\sigma, & r_{13} < \sigma, & r_{23} < \sigma \\
I^{(2)}: \sigma < r_{12} < g\sigma, & \sigma < r_{13} < g\sigma, & r_{23} < \sigma \\
I^{(3)}: \sigma < r_{12} < g\sigma, & \sigma < r_{13} < g\sigma, & \sigma < r_{23} < g\sigma.
\end{array}
$$

r_{12} refers to the distance between the centres of molecule 1 and molecule 2.

To evaluate these integrals we generally consider $\iint d\mathbf{r}_2 d\mathbf{r}_3$ integrated over the domains,

$$r_{12} < s, \quad r_{13} < t, \quad r_{23} < u,$$

and write it as $W(s,t,u)$. Since W is symmetric with respect to s, t, and u, it is sufficient to calculate only the case of $s \geqslant t$, and $s \geqslant u$. Obviously,

$$W(s, t, u) = 4\pi \int_0^u V(s, t, r) r^2 dr,$$

where $V(s, t, r)$ is the volume of the overlapping domain when a sphere of radius s and a sphere of radius t are located at a distance r between their centres, that is,

$$
\begin{aligned}
V(s, t, r) &= (\pi/12)[r^3 - 6r(s^2 + t^2) + 8(s^3 + t^3) - 3(s^2 - t^2)^2 r^{-1}] && (\text{for } s - t \leqslant r \leqslant s + t), \\
&= (4\pi/3)t^3 && (\text{for } r \leqslant s - t).
\end{aligned}
$$

Consequently,

$$
\begin{aligned}
W(s, t, u) &= (\pi^2/18)[s^6 + t^6 + u^6 + 18s^2t^2u^2 + 16(s^3t^3 + s^3u^3 + t^3u^3) \\
&\quad - 9\{s^4(t^2 + u^2) + t^4(s^2 + u^2) + u^4(s^2 + t^2)\}] \quad (\text{for } s \leqslant t + u), \\
&= (16\pi^2/9)t^3u^3 \quad (\text{for } s \geqslant t + u).
\end{aligned}
$$

With the aid of this function W, the integrals, $I^{(0)}, \ldots, I^{(3)}$ are expressed in the following form:

$$
\begin{aligned}
I^{(0)} &= W(\sigma, \sigma, \sigma) \\
I^{(1)} &= W(g\sigma, \sigma, \sigma) - I^{(0)} \\
I^{(2)} &= W(g\sigma, g\sigma, \sigma) - I^{(0)} - 2I^{(1)} \\
I^{(3)} &= W(g\sigma, g\sigma, g\sigma) - I^{(0)} - 3I^{(1)} - 3I^{(2)}.
\end{aligned}
$$

Finally, the third virial coefficient is

$$
\begin{aligned}
C(T) &= 2v_0^2[5 - (g^6 - 18g^4 + 32g^3 - 15)x \\
&\quad + (-2g^6 + 36g^4 - 32g^3 - 18g^2 + 16)x^2 \\
&\quad - (6g^6 - 18g^4 + 18g^2 - 6)x^3] \quad (g \leqslant 2) \\
&= 2v_0^2[5 - 17x + (32g^3 - 18g^2 - 48)x^2 \\
&\quad - (5g^6 - 32g^3 + 18g^2 + 26)x^3] \quad (g \geqslant 2)
\end{aligned}
$$

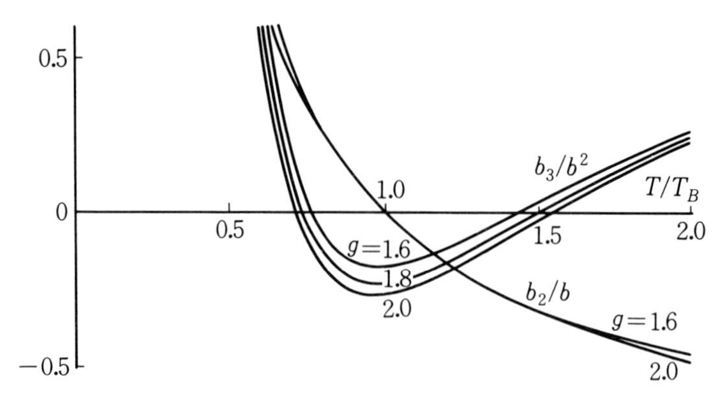

Figure 5.5 Dimensionless two-molecule and three-molecule cluster integrals for the spherically symmetric square-well potential. Note a close correspondence to Figure 5.2

(T. Kihara (1943), *Nippon Sugaku-Buturigaku-kaisi*, **17**, 11). For example,

$$C(T) = 2v_0^2(5 - 17x + 136x^2 - 162x^3) \quad (\text{when } g = 2).$$

Dimensionless values, $b_2^* = b_2/b$, $b_3^* = b_3/b^2$, of two-molecule and three-molecule cluster integrals,

$$b_2 = -B(T), \quad b_3 = 2B(T)^2 - C(T)/2,$$

are shown in Figure 5.5 as functions of T/T_B. This figure well reproduces the

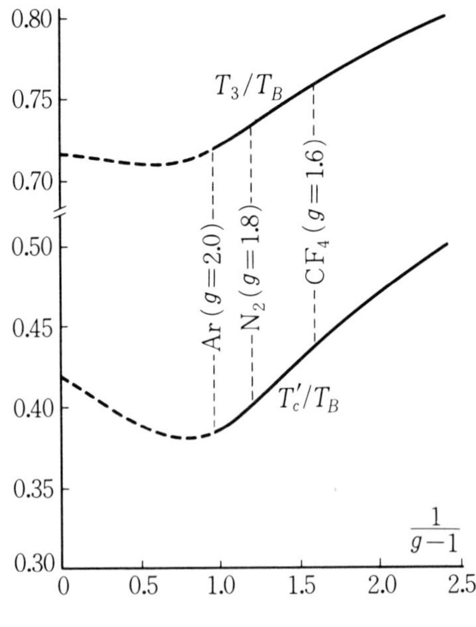

Figure 5.6

salient features of Figure 5.2 in which the measured values for argon, nitrogen, and carbon tetrafluoride are plotted; as one proceeds from monatomic molecules, diatomic molecules, to polyatomic molecules, existence of a hardcore in the molecule becomes pronounced and the value of g decreases from 2.0, 1.8, to 1.6.

Figure 5.6 depicts the ratio between the minimum zero T_3 of $b_3(T)$ and the Boyle temperature T_B as a function of $(g-1)^{-1}$. As one would expect, the portion of the solid line ($g \leqslant 2.0$) tends to agree with T_c/T_B in Table 5.1. In Figure 5.6 we also plot the ratio between the approximate value T_c' determined from the relation (5.21) and T_B. The latter agrees with the actual values within errors of the order of 10%.

Adopting a spherically symmetric square-well potential, we can thus express the second and third virial coefficients in terms of elementary functions. While this offers a great advantage, it is not suitable for an accurate treatment of intermolecular forces. In Chapter 6 we use smooth intermolecular potentials for rare gases; in Chapter 7 we discuss ways to take account of the molecular shapes for polyatomic molecules.

Chapter 6

The Lennard-Jones Potential

6.1 The second virial coefficients

In this chapter and the next, we shall treat the intermolecular forces in gases from the following point of view. On the basis of an appropriate intermolecular-potential function containing a few (usually two) constants, we determine those constants with the aid of measured values of the second virial coefficients; in the light of the intermolecular potential thus determined, we then examine the third virial coefficients or the three-molecule cluster integrals. In this chapter the rare-gas atoms are considered; in Chapter 7, the polyatomic molecules.

For the potential function taken initially as the basis, it is desirable to satisfy the following three conditions: proximity to real molecules (accuracy); applicability to various molecules (generality); and relative ease in carrying out the necessary integrations (integrability). These three conditions tend to be mutually incompatible. When the integrability is stressed, the situation tends to deviate far away from real molecules, as in the case of the square-well potential; an extremely accurate potential cannot be used for varieties of molecules. One which satisfies those three conditions most harmoniously for the rare-gas atoms is the Lennard-Jones potential; for the polyatomic molecules, it is the core potential.

The Lennard-Jones potential most conventionally used is a function of the following form:

$$U(r) = \lambda r^{-n} - \mu r^{-6} \qquad \lambda > 0,\ \mu > 0,\ n > 6.$$

Here r is the distance between the centres; the attractive part proportional to r^{-6} is chosen so as to conform with the form of the potential of the dispersion force. The power index n of the repulsive part is usually chosen to be 12. The reason for this choice first of all is the desirability that the two indices have a simple ratio for calculational convenience. Secondly, after the potential had been applied for various properties of rare gases, it was apparent that 9 is too small and 15 is too large for the index.

Choosing $n = 12$ and introducing U_0 and r_0 via

$$U(r) = \lambda r^{-12} - \mu r^{-6} = U_0\left[\left(\frac{r_0}{r}\right)^{12} - 2\left(\frac{r_0}{r}\right)^{6}\right], \tag{6.1}$$

the potential takes on its minimum value $-U_0$ at $r = r_0$. For rare-gas atoms of the same kind, r_0 corresponds to the atomic diameter. (Figure 1.2 shown earlier in this book depicts such a $U(r)$ accurately.)

The integration of the second virial coefficient,

$$B(T) = 2\pi \int_0^\infty (1 - e^{-U(r)/kT}) r^2 dr, \tag{6.2}$$

was carried out analytically by Lennard-Jones (J. E. Lennard-Jones (1924), *Proc. Roy. Soc., London*, A **106**, 463; J. E. Lennard-Jones (1931), *Proc. Phys. Soc., London, A* **43**, 461). After a partial integration it may be rewritten as

$$B(T) = -\frac{2\pi}{3} \int_0^\infty \frac{1}{kT} \frac{dU}{dr} e^{-U(r)/kT} r^3 dr. \tag{6.3}$$

Carrying out a power-series expansion of $\exp(\mu/r^6 kT)$ contained in $\exp\{-U(r)/kT\}$, one integrates the resulting expression term by term. The result can be expressed in a single power series,

$$B(T) = -\frac{\pi}{6}\left(\frac{\lambda}{kT}\right)^{1/4} \sum_{t=0}^{\infty} \Gamma\left(\frac{2t-1}{4}\right) \frac{y^t}{t!},$$
$$y \equiv (\mu/kT)(kT/\lambda)^{1/2} = 2(U_0/kT)^{1/2}. \tag{6.4}$$

Here Γ is the gamma function; the coefficients can be evaluated from $\Gamma(-1/4) = -4.9017$, $\Gamma(1/4) = 3.6256$, and the formula $\Gamma(x+1) = x\Gamma(x)$.

Expressing (6.4) in terms of U_0 and r_0 in (6.1), we have

$$B(T) = -\frac{\pi}{6} r_0^3 \left(\frac{U_0}{kT}\right)^{1/4} \sum_{t=0}^{\infty} \frac{2^t}{t!} \Gamma\left(\frac{2t-1}{4}\right)\left(\frac{U_0}{kT}\right)^{t/2}. \tag{6.5}$$

In these formulas we see that $B(T)$ is positive at high temperatures and negative at low temperatures. Quantities related to the Boyle temperatures are calculated as

$$kT_B = 3.418 U_0, \quad b \equiv (dB/d \ln T)_{T=T_B} = 1.201 r_0^3.$$

When the measured values of the second virial coefficients are available over a sufficiently wide range of temperatures, one can determine the constants, U_0 and r_0, through examination of the potential with the aid of (6.5). We may, for example, proceed in the following way. Setting (6.5) as

$$B(T) = (2\pi/3) r_0^3 \, F(z), \quad z \equiv U_0/kT, \tag{6.6}$$

we plot $\log |F(z)|$ on the ordinate and $-\log z$ on the abscissa. Separately, we plot $\log |B(T)|$ on the ordinate and $\log T$ on the abscissa for measured values. If the potential function accurately represents the reality and the errors involved in the measurement are small, those two graphs should overlap by parallel translations. The magnitudes of $2\pi r_0^3/3$ and U_0/k are then determined from the distances of the coordinate axes between the two. For the numerical values of $F(z)$, Table 7.2 (see Section 7.4) can be used.

In fact for rare gases the agreement is almost perfect; the constants so

Table 6.1 The potential constants

	U_0/k (K)	r_0 (Å)	$2r_0^6 U_0$ (atomic units)
He	10.3	2.91	1.80
Ne	35.8	3.09	8.99
Ar	119	3.83	108
Kr	167	4.13	239
Xe	225	4.57	591

The values for Kr and Xe are taken from E. Whalley and W. G. Schneider (1955). *J. Chem. Phys.*, **23**, 1644. The rest are from T. Kihara (1958). *Adv. Chem. Phys.*, Vol. 1.

determined are listed in Table 6.1. The values for helium here are determined with the aid of the expression (Section 6.4) which takes account of the quantum-mechanical effects.

The values $2r_0^6 U_0$ of the coefficient of r^{-6} are considerably greater than the values of the coefficient of the potential $-\mu r^{-6}$ for the dispersion force listed in Section 4.3:

He 1.47, Ne 6.3, Ar 65, Kr 130, Xe 270.

The reason may be traced to the fact that the second term $-2r_0^6 U_0/r^6$ of the Lennard-Jones potential contains not only the potential of the dispersion force of the dipole–dipole type but also implicitly the interaction of the dipole–quadrupole type proportional to r^{-8}.

6.2 Three-molecule cluster integrals

Adoption of the Lennard-Jones function $U(r)$ given by (6.1) as the potential between two molecules leads to adoption of the function

$$U(r_{12}, r_{23}, r_{31}) \equiv U(r_{12}) + U(r_{23}) + U(r_{31}) + u(r_{12}, r_{23}, r_{31}) \tag{6.7}$$

as the potential between three similar molecules. Here r_{ij} is the distance between two molecules, i and j, and $u(r_{12}, r_{23}, r_{31})$ represents an extra term arising from nonadditivity of the potential. An expression for this term with a least amount of arbitrariness may be obtained by using its asymptotic form:

$$u(r_{12}, r_{23}, r_{31}) = \nu(r_{12}r_{23}r_{31})^{-3}(3 \cos\theta_1 \cos\theta_2 \cos\theta_3 + 1), \tag{6.8}$$

where θ_1, θ_2, and θ_3 are the angles of a triangle formed by the three molecules; the coefficient ν has been given in Section 4.4. The purpose of this section is to analyse the three-molecule cluster integrals based on such a three-molecule potential.

The expression for the three-molecule cluster integral b_3 has been given in

Section 5.6, that is,

$$b_3 = \frac{4\pi^2}{3} \iiint \left[\exp\left(\frac{-U(r_{12}, r_{23}, r_{31})}{kT}\right) - \exp\left(\frac{-U(r_{12})}{kT}\right) - \exp\left(\frac{-U(r_{23})}{kT}\right) - \exp\left(\frac{-U(r_{31})}{kT}\right) + 2 \right] r_{12}r_{23}r_{31}dr_{12}dr_{23}dr_{31}, \tag{6.9}$$

where the integration variables have been transformed from $d\mathbf{r}_2 d\mathbf{r}_3$ to $8\pi^2 r_{12}r_{23}r_{31}dr_{12}dr_{23}dr_{31}$. The integrals are to be carried out over all values of r_{12}, r_{23}, and r_{31} forming three sides of a triangle.

Problem

Derive the transformation of integration variables stated above.

Solution

Choosing the origin at molecule 1 and the z axis along the line connecting molecule 1 and molecule 2, we have

$$d\mathbf{r}_2 d\mathbf{r}_3 = 4\pi r_{12}^2 \, dr_{12} 2\pi r dr dz,$$

where r is the distance of the molecule 3 from the z axis; z is its z coordinate. Expressing (r_{13}, r_{23}) in terms of (r,z), we have

$$r_{13}^2 = z^2 + r^2, \quad r_{23}^2 = (z - r_{12})^2 + r^2.$$

The Jacobian is

$$\frac{\partial(r_{13}^2, r_{23}^2)}{\partial(z, r^2)} = \begin{vmatrix} 2z & 1 \\ 2(z - r_{12}) & 1 \end{vmatrix} = 2r_{12},$$

whence

$$r_{13}r_{23}dr_{13}dr_{23} = r_{12}rdrdz.$$

Consequently

$$d\mathbf{r}_2 d\mathbf{r}_3 = 8\pi^2 r_{12}r_{23}r_{31}dr_{12}dr_{23}dr_{21}.$$

Now, since the effect of nonadditivity is rather small, we may write

$$b_3 = (b_3)_{\nu=0} + \nu(\partial b_3/\partial\nu)_{\nu=0}, \tag{6.10}$$

where

$$(b_3)_{\nu=0} = \frac{4\pi^2}{3} \iiint [f(r_{12}, r_{23}, r_{31}) - f(r_{12}) - f(r_{23}) - f(r_{31})] \times r_{12}r_{23}r_{31}dr_{12}dr_{23}dr_{31},$$

$$f(r) = \exp[-U(r)/kT] - 1,$$

$$f(r_{12}, r_{23}, r_{31}) = \exp\left[-\frac{U(r_{12}) + U(r_{23}) + U(r_{31})}{kT}\right] - 1,$$

$$\left(\frac{\partial b_3}{\partial \nu}\right)_{\nu=0} = -\frac{4\pi^2}{3}\frac{1}{kT}\iiint \exp\left[-\frac{U(r_{12}) + U(r_{23}) + U(r_{31})}{kT}\right]$$

$$\times \frac{3\cos\theta_1\cos\theta_2\cos\theta_3 + 1}{(r_{12}r_{23}r_{31})^2} dr_{12}dr_{23}dr_{31}.$$

Those integrals are equal to three times the integrals in which r_{12} is taken to be the maximum of r_{12}, r_{23}, and r_{31}. Choosing new integration variables,

$$R \equiv r_{12}, \quad \xi \equiv r_{13}/R, \quad \eta \equiv r_{23}/R,$$

we may rewrite $(b_3)_{\nu=0}$ as

$$4\pi^2 \int_0^1 \int_{1-\xi}^1 \left[\int_0^\infty \{f(R, \xi R, \eta R) - f(R) - f(\xi R) - f(\eta R)\}R^5\, dR\right] \xi\eta d\eta d\xi.$$

To carry out the integration with respect to R, we may tentatively set

$$U(r) = \lambda r^{-12} - \mu r^{-m}, \quad m > 6$$

and then take the limit of $m \to 6$ (T. Kihara (1948), *J. Phys. Soc., Japan*, **3**, 265). For $m > 6$, we may integrate term by term as in the previous section; the result is

$$\int_0^\infty \{f(R, \xi R, \eta R) - f(R) - f(\xi R) - f(\eta R)\}R^5\, dR = \left(\frac{\lambda}{kT}\right)^{1/2} \sum_{t=0}^{\infty} C_t y^t,$$

$$C_t = \frac{1}{12}\frac{1}{t!}\Gamma\left(\frac{tm-6}{12}\right)[\langle m\rangle^t \langle 12\rangle^{(6-mt)/12} - \langle 6\rangle],$$

where the symbol $\langle \ldots \rangle$ is

$$\langle \alpha \rangle \equiv 1 + \xi^{-\alpha} + \eta^{-\alpha}.$$

In the limit of $m \to 6$,

$$C_t = \frac{1}{12}\frac{1}{t!}\Gamma\left(\frac{t-1}{2}\right)[\langle 6\rangle^t \langle 12\rangle^{(1-t)/2} - \langle 6\rangle] \quad (t \neq 1)$$

$$C_1 = \frac{1}{12}[\xi^{-6}\ln\xi^{-12} + \eta^{-6}\ln\eta^{-12} - \langle 6\rangle \ln\langle 12\rangle].$$

We thus obtain the following result in a form of power series analogous to (6.4):

$$(b_3)_{\nu=0} = \left(\frac{\lambda}{kT}\right)^{1/2} \sum_{t=0}^{\infty} G_t y^t, \tag{6.11}$$

$$G_t = 4\pi^2 \int_0^1 \int_{1-\xi}^1 C_t \xi\eta d\eta d\xi.$$

Table 6.2 Coefficients for equations 6.11, 6.12, and 6.26

t	Gt	Ht	It
0	11.278	2.350	+211.8
1	−17.733	1.080	−80.22
2	+3.498	0.557	−40.42
3	1.492	0.268	−33.80
4	0.571	0.1203	−20.55
5	0.201	0.0505	−9.85
6	0.0662	0.0201	−4.75
7	0.0206	0.0076	−1.652
8	0.0062	0.0027	−0.586
9	0.00178		−0.197
10	0.00049		−0.063
11	0.00014		

From T. Kihara (1958). *Adv. Chem. Phys.*, Vol. 1.

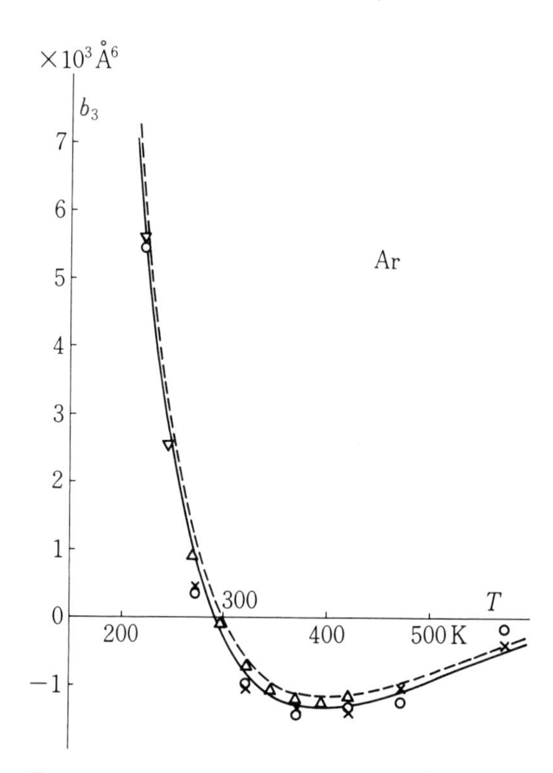

Figure 6.1 Three-molecule cluster integral for argon. The solid line represents the calculated values taking account of nonadditivity in the intermolecular potential; the dashed line, those neglecting nonadditivity. ○, x, and △ distinguish the different experimentalists

The values of G_t obtained by means of numerical integration are listed in Table 6.2.

Correction arising from nonadditivity of the potential is similarly obtained as

$$\left(\frac{\partial b_3}{\partial \nu}\right)_{\nu=0} = -\frac{1}{kT}\left(\frac{kT}{\lambda}\right)^{1/4} \sum_{t=0}^{\infty} H_t y^t \tag{6.12}$$

$$H_t = \frac{\pi^2}{3}\frac{1}{t!}\Gamma\left(\frac{2t+1}{4}\right)\int_0^1 \int_{1-\xi}^1 \langle 6 \rangle^t \langle 12 \rangle^{-(2t+1)/4}$$

$$\times \frac{3\cos\theta_1 \cos\theta_2 \cos\theta_3 + 1}{\xi^2 \eta^2} d\eta d\xi.$$

The numerical values of H_t are also listed in Table 6.2. We remark that H_t has the same sign for all the values of $t = 0, 1, 2, \ldots$.

The computed values and measured values of the three-molecule cluster integral b_3 are compared in Figures 6.1 and 6.2 for argon and krypton, respectively. For the computation, the potential constants, U_0 and r_0, in Table 6.1 are used. The dashed line represents the result in which nonadditivity (6.8) of the potential is neglected; the solid line takes account of it. The measured values of b_3 are obtained from

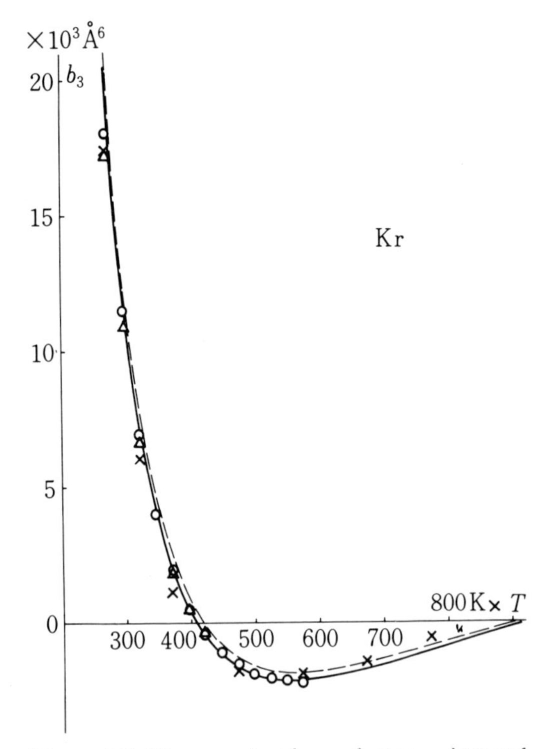

Figure 6.2 Three-molecule cluster integral for krypton. The solid and dashed lines have the same meaning as in figure 6.1

those of the second and third virial coefficients with the aid of the relation $b_3 = 2B^2 - C/2$. It is clear that a satisfactory agreement is obtained by taking account of nonadditivity.

As may be clear from the figures, it may be concluded that the effect of nonadditivity is such as to shift the theoretical curve of b_3 downward. In Chapter 7 these aspects will be examined qualitatively for polyatomic molecules.

6.3 Four-molecule and five-molecule cluster integrals

Recent advances in electronic computers have made it possible to carry out computations of higher-order cluster integrals. Usually the computation is carried out in the form of virial coefficients; for an additive Lennard-Jones potential, up to the fifth virial coefficient have been obtained (J. A. Baker, P. J. Leonard, and A. Pompe (1966), *J. Chem. Phys.*, **44**, 4206). In terms of dimensionless quantities,

$$B^* \equiv B/b \quad C^* \equiv C/b^2, \quad D^* \equiv D/b^3, \quad E^* \equiv E/b^4,$$

Figure 6.3 has been plotted.

The solid lines in Figure 6.4 express the fourth and fifth cluster integrals in terms of dimensionless quantities, $b_4^* \equiv b_4/b^3$ and $b_5^* \equiv b_5/b^4$. The dashed lines are those obtained from (5.4) by setting $D(T)$ and $E(T)$ equal to zero. We see that the contributions of $D(T)$ and $E(T)$ are small at or below the Boyle temperature.

The foregoing have been concerned with an additive Lennard-Jones potential. Nonadditivity affects all of $C(T)$, $D(T)$, and $E(T)$ to a certain degree; based on the

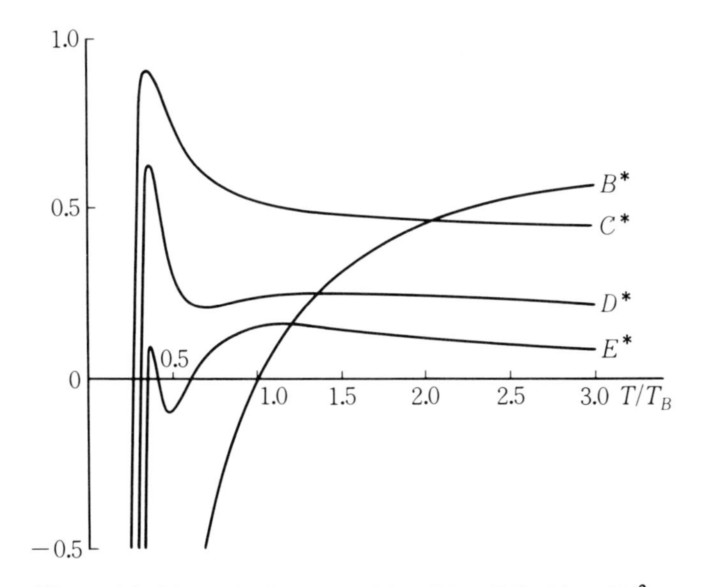

Figure 6.3 Dimensionless quantities, $B^* \equiv B/b$, $C^* \equiv C/b^2$, . . . , reduced from the second to fifth virial coefficients with respect to additive Lennard-Jones potential

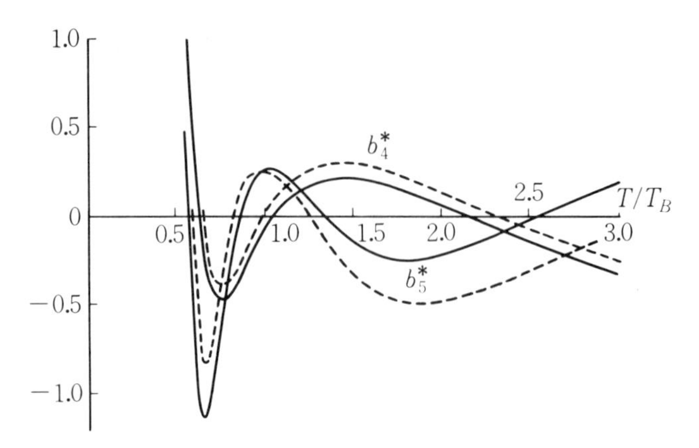

Figure 6.4 The solid lines describe the four-molecule and the five-molecule cluster integrals computed from Figure 6.3; the dashed lines correspond to those obtained by setting D^* and E^* equal to zero

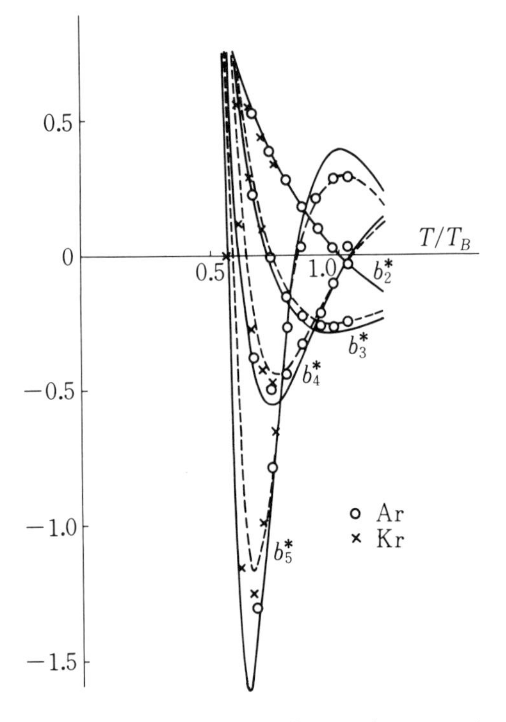

Figure 6.5 The two-, three-, four-, and five-molecule cluster integrals for argon and krypton. The solid lines take account of corrections arising from nonadditivity in the intermolecular potential; the dashed lines neglect such corrections

earlier statement, it would be sufficient to consider the effect arising only through $C(T)$ at or below the Boyle temperature. For calculation of such an effect, the results of the previous section may be used.

The computed curves for argon and krypton are shown in Figure 6.5; they are compared with the measured values taken from Figure 5.1. The solid lines take account of nonadditivity of the three-body force as stated above; the dashed lines neglect such an effect. (The effect of nonadditivity can be treated almost commonly for argon and krypton; hence only the case of argon has been considered.) Considering the errors involved in the measured values we may regard the result as satisfactory.

6.4 Quantum corrections to the cluster integrals

As is well known, the de Broglie wavelength associated with a particle with a momentum p is $2\pi\hbar/p$. When this wavelength becomes comparable to the particle diameter, the effect of diffraction appears in the event of interparticle collision. When one deals with cluster integrals, such a quantum effect must be taken into consideration for light and small molecules such as helium and hydrogen. (The quantum effect arising from symmetry properties of wave functions with respect to interchange of molecules need not be considered except in the case of helium at temperatures below 25 K.)

As we have seen in Section 5.6, the cluster integral is written generally as integration of a linear combination of the form

$$\exp[-\beta\Phi(\mathbf{r}_1\ \mathbf{r}_2, \ldots, \mathbf{r}_l)], \quad \beta \equiv 1/kT, \tag{6.13}$$

Here $\Phi(\mathbf{r}_1, \ldots, \mathbf{r}_l)$ is the potential energy of an l-molecule system, and $\mathbf{r}_1, \ldots, \mathbf{r}_l$ represent the positions of molecules. The Boltzmann factor (6.13) represents the probability density that the l-molecule system occupies a point $\mathbf{r}_1, \ldots, \mathbf{r}_l$ in the configuration space in the canonical distribution. It is normalized so that it takes on unity when l molecules are sufficiently far apart from each other. The purpose of this section is to generalize the Boltzmann factor so that the quantum effects may be taken into consideration. Specifically we consider an expansion with respect to h^2 in which (6.13) represents the first term, and thereby calculate the successive terms proportional to $\hbar^2$, $\hbar^4$, etc.

Considering the Hamiltonian for the l-molecule system,

$$H = -\frac{\hbar^2}{2m}\sum_{i=1}^{l}\Delta_i + \Phi(\mathbf{r}_1, \ldots, \mathbf{r}_l) \qquad \Delta_i \equiv \nabla_i \cdot \nabla_i, \quad \nabla_i \equiv \partial/\partial\mathbf{r}_i \tag{6.14}$$

(m is the mass of a molecule), we denote by ψ_ν ([$\mathbf{r}_1, \ldots, \mathbf{r}_l$) the eigenfunction associated with the eigenvalue E_ν:

$$H\psi_\nu(\mathbf{r}_1, \ldots, \mathbf{r}_l) = E_\nu\psi_\nu(\mathbf{r}_1, \ldots, \mathbf{r}_l). \tag{6.15}$$

The function ψ_ν is normalized so that

$$\int \ldots \int \psi_\nu(\mathbf{r}_1, \ldots, \mathbf{r}_l)\psi_{\nu'}{}^*(\mathbf{r}_1, \ldots, \mathbf{r}_l)\, d\mathbf{r}_1 \ldots d\mathbf{r}_l = \delta_{\nu\nu'}.$$

An arbitrary function $\psi(\mathbf{r}_1, \ldots, \mathbf{r}_l)$ may be expanded in terms of the eigenfunctions ψ_ν as

$$\psi(\mathbf{r}_1, \ldots, \mathbf{r}_l) = \Sigma\, a_\nu \psi_\nu(\mathbf{r}_1, \ldots, \mathbf{r}_l),$$

where

$$a_\nu = \int \ldots \int \psi(\mathbf{r}_1', \ldots, \mathbf{r}_l')\psi_\nu{}^*(\mathbf{r}_1', \ldots, \mathbf{r}_1')d\mathbf{r}_1' \ldots d\mathbf{r}_l'.$$

Substituting this expression for a_ν in the foregoing formula, we have

$$\psi(\mathbf{r}_1, \ldots, \mathbf{r}_l) = \int \ldots \int \psi(\mathbf{r}_1', \ldots, \mathbf{r}_l')\, \Sigma\, \psi_\nu(\mathbf{r}_1, \ldots, \mathbf{r}_l) \times \psi_\nu{}^*(\mathbf{r}_1', \ldots, \mathbf{r}_l')d\mathbf{r}_1' \ldots d\mathbf{r}_l'.$$

With the aid of the Dirac δ-function, $\delta(\mathbf{r} - \mathbf{r}')$, we thus obtain

$$\sum_\nu \psi_\nu(\mathbf{r}_1, \ldots, \mathbf{r}_l)\psi_\nu{}^*(\mathbf{r}_1', \ldots, \mathbf{r}_l') = \delta(\mathbf{r}_1 - \mathbf{r}_1') \ldots \delta(\mathbf{r}_l - \mathbf{r}_l'). \tag{6.16}$$

The density matrix for the l-molecule system may be defined as

$$\rho(\mathbf{r}_1, \ldots, \mathbf{r}_l; \mathbf{r}_1', \ldots, \mathbf{r}_l') = \sum_\nu \exp(-\beta E_\nu)\psi_\nu(\mathbf{r}_1, \ldots, \mathbf{r}_l)\psi_\nu{}^*(\mathbf{r}_1', \ldots, \mathbf{r}_l'). \tag{6.17}$$

Its diagonal element, $\rho(\mathbf{r}_1, \ldots, \mathbf{r}_l; \mathbf{r}_1, \ldots, \mathbf{r}_l)$, is proportional to the probability density that the l-molecule system occupies a point, $\mathbf{r}_1, \ldots, \mathbf{r}_l$, in the configuration space in the canonical distribution.

The density matrix (6.17) satisfies the Bloch differential equation,

$$\frac{\partial}{\partial \beta}\rho(\mathbf{r}_1, \ldots, \mathbf{r}_l; \mathbf{r}_1', \ldots, \mathbf{r}_l') = -H\rho(\mathbf{r}_1, \ldots, \mathbf{r}_l; \mathbf{r}_1', \ldots, \mathbf{r}_l'), \tag{6.18}$$

where the operator H is to operate only on the variables, $\mathbf{r}_1, \ldots, \mathbf{r}_l$, without a prime in the density matrix. The 'initial condition' for this differential equation is

$$\lim_{\beta \to 0} \rho(\mathbf{r}_1, \ldots, \mathbf{r}_l; \mathbf{r}_1', \ldots, r_l') = \delta(\mathbf{r}_1 - \mathbf{r}_1') \ldots \delta(\mathbf{r}_l - \mathbf{r}_l'), \tag{6.19}$$

obtained from (6.16).

In a special case that the potential Φ is zero, the solution to (6.18) satisfying (6.19) is

$$\rho(\mathbf{r}_1, \ldots, \mathbf{r}_l; \mathbf{r}_1', \ldots, \mathbf{r}_l') = \left(\frac{m}{2\pi\hbar^2\beta}\right)^{3l/2} \exp\left[-\frac{m}{2\hbar^2\beta}\sum_i (\mathbf{r}_i - \mathbf{r}_i')^2\right].$$

As a general solution for $\Phi \neq 0$, we may thus anticipate the following form,

$$\rho(\mathbf{r}_1, \ldots, \mathbf{r}_l; \mathbf{r}_1', \ldots, \mathbf{r}_l') = \left(\frac{m}{2\pi\hbar^2\beta}\right)^{3l/2} \exp\left[-\frac{m}{2\hbar^2\beta}\sum_i (\mathbf{r}_i - \mathbf{r}_i')^2 + \Psi(\mathbf{r}_1, \ldots, \mathbf{r}_l; \mathbf{r}_1', \ldots, \mathbf{r}_l')\right], \tag{6.20}$$

$$\Psi \to 0 \quad \text{as} \quad \beta\Phi \to 0. \tag{6.21}$$

The product between the diagonal element of the density matrix and $(2\pi\hbar^2\beta/m)^{3l/2}$ is

$$(2\pi\hbar^2\beta/m)^{3l/2}\rho(\mathbf{r}_1, \ldots, \mathbf{r}_l; \mathbf{r}_1, \ldots, \mathbf{r}_l) = \exp \Psi(\mathbf{r}_1, \ldots, \mathbf{r}_l; \mathbf{r}_1, \ldots, \mathbf{r}_l). \tag{6.22}$$

That either side of this equation represents an expression for the Boltzmann factor with quantum effects taken into account may be understood from the fact that this quantity describes the probability density of occupying $\mathbf{r}_1, \ldots, \mathbf{r}_l$ in the configuration space and that it is normalized to unity at $\Phi = 0$.

Hereafter, let us write $\Psi(\mathbf{r}_1, \ldots, \mathbf{r}_l; \mathbf{r}_l; \mathbf{r}'_1, \ldots, \mathbf{r}'_l)$ simply as Ψ. Substitution of (6.20) in the Bloch equation yields

$$\Phi + \frac{\partial \Psi}{\partial \beta} + \frac{1}{\beta}\sum_i (\mathbf{r}_i - \mathbf{r}'_i) \cdot \nabla_i \Psi = \frac{\hbar^2}{2m} \sum_i [(\nabla_i \Psi)^2 + \Delta_i \Psi].$$

We now expand Ψ in a power series of $\hbar^2/2m$:

$$\Psi = \Psi_0 + (\hbar^2/2m)\Psi_1 + (\hbar^2/2m)^2\Psi_2 + \ldots. \tag{6.23}$$

Paying attention to the terms proportional to the zeroth, the first, the second, ... powers of $\hbar^2/2m$, we obtain

$$\Phi + \frac{\partial \Psi_0}{\partial \beta} + \frac{1}{\beta}\sum_i (\mathbf{r}_i - \mathbf{r}'_i) \cdot \nabla_i \Psi_0 = 0,$$

$$\frac{\partial \Psi_1}{\partial \beta} + \frac{1}{\beta}\sum_i (\mathbf{r}_i - \mathbf{r}'_i) \cdot \nabla_i \Psi_1 = \sum_i (\nabla_i \Psi_0)^2 + \sum_i \Delta_i \Psi_0,$$

$$\frac{\partial \Psi_2}{\partial \beta} + \frac{1}{\beta}\sum_i (\mathbf{r}_i - \mathbf{r}'_i) \cdot \nabla_i \Psi_2 = 2\sum_i \nabla_i \Psi_0 \cdot \nabla_i \Psi_1 + \sum_i \Delta_i \Psi_1.$$

On account of (6.21), the first equation above yields an expected result,

$$\Psi_0(\mathbf{r}_1, \ldots, \mathbf{r}_l; \mathbf{r}_1, \ldots, \mathbf{r}_l) = -\beta\Phi(\mathbf{r}_1, \ldots, \mathbf{r}_l).$$

Operating ∇_i and Δ_i onto the first equation we then obtain

$$\nabla_i \Phi + \frac{\partial}{\partial \beta} \nabla_i \Psi_0 + \frac{1}{\beta} \nabla_i \Psi_0 + \frac{1}{\beta} \sum_j (\mathbf{r}_j - \mathbf{r}'_j) \cdot \nabla_j \nabla_i \Psi_0 = 0,$$

$$\Delta_i \Phi + \frac{\partial}{\partial \beta} \Delta_i \Psi_0 + \frac{2}{\beta} \Delta_i \Psi_0 + \frac{1}{\beta} \sum_j (\mathbf{r}_j - \mathbf{r}'_j) \cdot \nabla_j \Delta_i \Psi_0 = 0.$$

The limits of $\mathbf{r}' \to \mathbf{r}$ are

$$(\nabla_i \Psi_0)_{\mathbf{r}' \to \mathbf{r}} = -\tfrac{1}{2}\beta \nabla_i \Phi, \quad (\Delta_i \Psi_0)_{\mathbf{r}' \to \mathbf{r}} = -\tfrac{1}{3}\beta \Delta_i \Phi.$$

Substituting these into the second equation above we obtain

$$\Psi_1(\mathbf{r}_1, \ldots, \mathbf{r}_l; \mathbf{r}_1, \ldots, \mathbf{r}_l) = \sum_i \left[\frac{\beta^3}{12} (\nabla_i \Phi)^2 - \frac{\beta^2}{6} \Delta_i \Phi \right].$$

Similarly,

$$\Psi_2(\mathbf{r}_1, \ldots, \mathbf{r}_l; \mathbf{r}_1, \ldots, \mathbf{r}_l)$$

$$= \sum_j \sum_i \left[-\frac{\beta^5}{60} \nabla_i \nabla_j \Phi : \nabla_i \Phi \nabla_j \Phi + \frac{\beta^4}{90} \nabla_i \nabla_j \Phi : \nabla_i \nabla_j \Phi \right.$$

$$\left. + \frac{\beta^4}{30} \nabla_i \Phi \cdot \nabla_i \Delta_j \Phi - \frac{\beta^3}{60} \Delta_i \Delta_j \Phi \right].$$

Here the two-dot symbol : implies a scalar product between two tensors.

Replacing the Boltzmann factor (6.13) by this exp $\Psi(\mathbf{r}_1, \ldots, \mathbf{r}_l; \mathbf{r}_1, \ldots, \mathbf{r}_l)$ in the integrand of the cluster integrals b_l, we obtain the expression including quantum effects in the form,

$$b_l = b_l^{(0)} + (\hbar^2 \beta/4m) b_l^{(1)} + (\hbar^2 \beta/4m)^2 b_l^{(2)} + \ldots \tag{6.24}$$

$(l = 2, 3, \ldots, \beta = 1/kT)$.

Account of quantum effects through such an expansion was first taken by Wigner (E. Wigner (1932), *Phys. Rev.*, **40**, 749) for the two-molecule cluster integral, b_2. The calculations in this section are mostly due to Kihara *et al.* (T. Kihara, Y. Midzuno, and T. Shizume (1955), *J. Phys. Soc., Japan*, **10**, 249); they also owe to a large extent to the paper by Husimi (K. Husimi (1940), *Proc. Phys. Math. Soc. Japan*, **22**, 264) concerning the density matrix.

The final results for the two-molecule cluster integral are

$$b_2^{(1)} = -\frac{2\pi}{3} \beta^2 \int_0^\infty U'^2 e^{-\beta U} r^2 dr,$$

$$b_2^{(2)} = \frac{8\pi}{3} \beta^2 \int_0^\infty \left(\frac{U''^2}{10} + \frac{U'^2}{5r^2} + \frac{\beta U'^3}{9r} - \frac{\beta^2 U'^4}{72} \right) e^{-\beta U} r^2 dr,$$

$$b_2^{(3)} = -64\pi\beta^2 \int_0^\infty \left(\frac{U'''^2}{840} + \frac{U''^2}{140r^2} + \frac{\beta U''^3}{756} + \frac{\beta U' U''^2}{180r} \right.$$

$$\left. + \frac{\beta U'^3}{945r^3} - \frac{\beta^2 U'^2 U''^2}{720} - \frac{\beta^2 U'^4}{6480r^2} - \frac{\beta^3 U'^5}{2160r} + \frac{\beta^4 U'^6}{25920} \right) e^{-\beta U} r^2 dr.$$

Here $U \equiv U(r) \equiv \Phi(\mathbf{r}_1, \mathbf{r}_2)$ are the intermolecular potentials; U', U'', etc. are derivatives of $U(r)$.

For the three-molecule cluster integral, setting $\Phi(\mathbf{r}_1, \mathbf{r}_2, \mathbf{r}_3)$ as $U(r_{12}, r_{23}, r_{31})$, one has

$$b_3^{(1)} = -\frac{4\pi^2}{9} \beta^2 \iiint (A_{12} + A_{23} + A_{31}) r_{12} r_{23} r_{31} dr_{12} dr_{23} dr_{31},$$

$$A_{12} = \left[\left(\frac{\partial U(r_{12}, r_{23}, r_{31})}{\partial r_{12}} \right)^2 + \cos\theta_3 \frac{\partial U(r_{12}, r_{23}, r_{31})}{\partial r_{13}} \right.$$

$$\left. \times \frac{\partial U(r_{12}, r_{23}, r_{31})}{\partial r_{23}} \right] \exp[-\beta U(r_{12}, r_{23}, r_{31})]$$

$$-[U'(r_{12})]^2 \exp[-\beta U(r_{12})];$$

A_{23} and A_{31} may be obtained similarly. θ_i is the corner angle at the molecule i for a triangle formed by the molecules 1, 2, and 3. The integration is carried out for all the r_{12}, r_{23}, and r_{31}, forming the triangle.

Now, adopting the Lennard-Jones potential,

$$U(r) = \lambda r^{-12} - \mu r^{-6} = U_0\left[\left(\frac{r_0}{r}\right)^{12} - 2\left(\frac{r_0}{r}\right)^{6}\right],$$

we may calculate the various terms of (6.24) for the two-molecule cluster integral b_2 in the following form:

$$b_2^{(1)} = -\frac{\pi}{18}\left(\frac{\lambda}{kT}\right)^{1/12}\sum_{t=0}^{\infty}(36t-11)\Gamma\left(\frac{6t-1}{12}\right)\frac{y^t}{t!},$$

$$b_2^{(2)} = \frac{\pi}{180}\left(\frac{\lambda}{kT}\right)^{-1/12}\sum_{t=0}^{\infty}(3024t^2+4728t+767)\Gamma\left(\frac{6t+1}{12}\right)\frac{y^t}{t!},$$

$$b_2^{(3)} = -\frac{\pi}{420}\left(\frac{\lambda}{kT}\right)^{-1/4}\sum_{t=0}^{\infty}(53568t^3+303216t^2$$

$$+491076t+180615)\Gamma\left(\frac{2t+1}{4}\right)\frac{y^t}{t!},$$

where

$$y \equiv (\mu/kT)(kT/\lambda)^{1/2} = 2(U_0/kT)^{1/2}.$$

On the basis of these expressions, we may determine the potential between helium atoms with the aid of the measured values of the second virial coefficient, $B(T) \equiv -b_2(T)$. The values listed in Table 6.1 are the results of such computations. To see the extent of the quantum effect, Figure 6.6 includes the classical values,

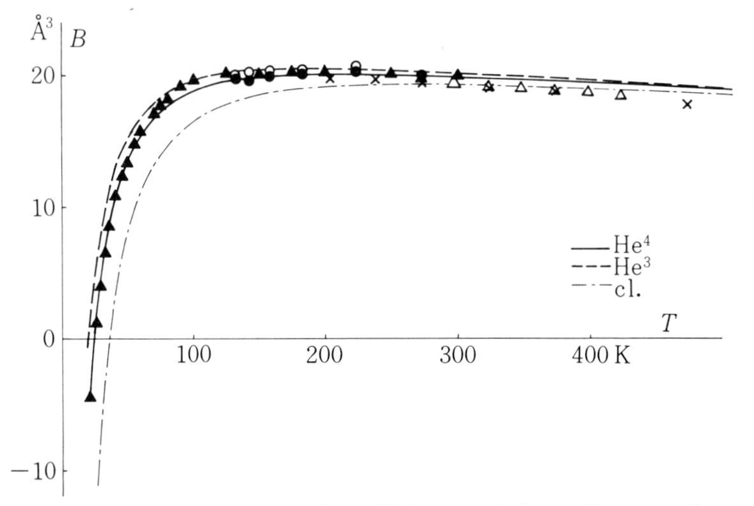

Figure 6.6 The second virial coefficient for helium. The chain line represents the calculated values without taking account of the quantum effects. Different marks on the measured values for He^4 distinguish the different experimentalists

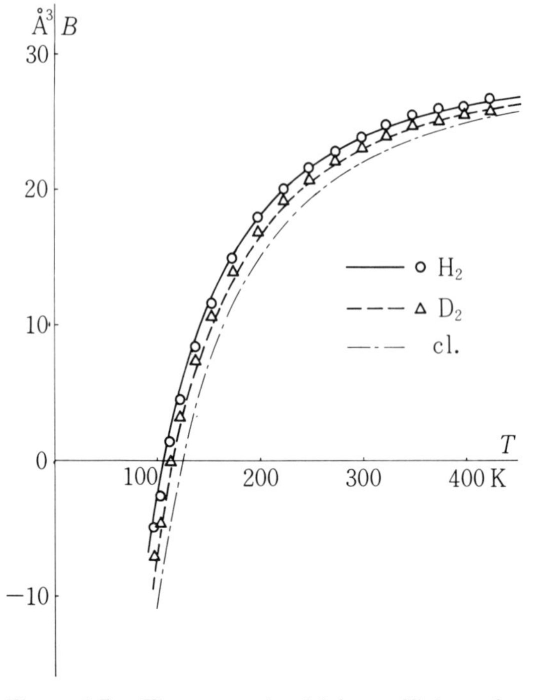

Figure 6.7 The second virial coefficient for hydrogen H_2 and deuterium D_2

$-b_0^{(0)}$. For He^3, since its interatomic potential should be the same as that of He^4, we may depict a similar curve through consideration of the difference in the mass m; such a result is also shown in the figure. (As we have remarked in the beginning of this section, the exchange effects are negligible in the temperature domain applicable to the cases of the figure.)

As may be seen in the figure, the measured values do not fit on the curve at high temperatures exceeding ten times the Boyle temperature. We may interpret this as an indication of the limit of accuracy for the Lennard-Jones potential.

Figure 6.7 depicts similar curves for hydrogen H_2 and deuterium D_2. By fitting with the measured values, the constants in the Lennard-Jones potential can be determined; the result is

$$U_0/k = 36.3 \text{ K}, \qquad r_0 = 3.28 \text{ Å}.$$

For the three-molecule cluster integral b_3,

$$b_3 \approx (b_3^{(0)})_{\nu=0} + \nu \left(\frac{\partial b_3^{(0)}}{\partial \nu}\right)_{\nu=0} + \frac{\hbar^2}{4mkT} (b_3^{(1)})_{\nu=0}. \tag{6.25}$$

The first and second terms on the right-hand side have been obtained in Section 6.2;

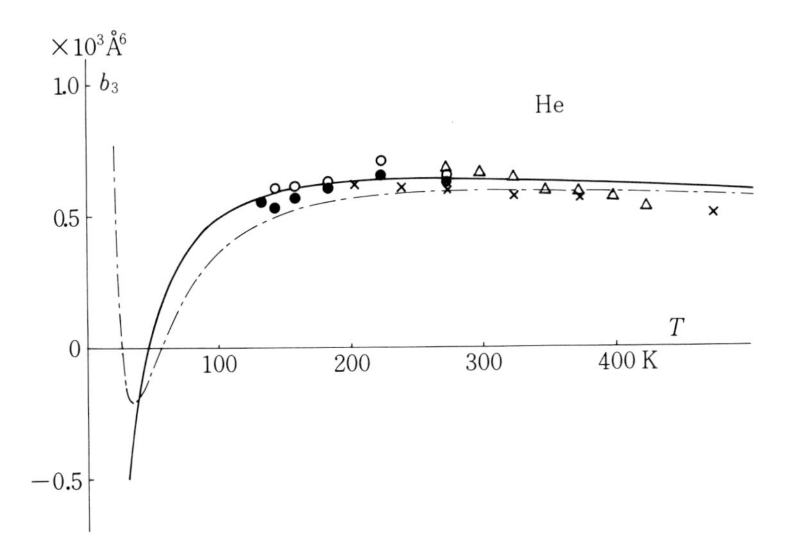

Figure 6.8 The three-molecule cluster integral for helium. The chain line represents the calculated values without taking account of the quantum effects. Different marks on the measured values distinguish the different experimentalists

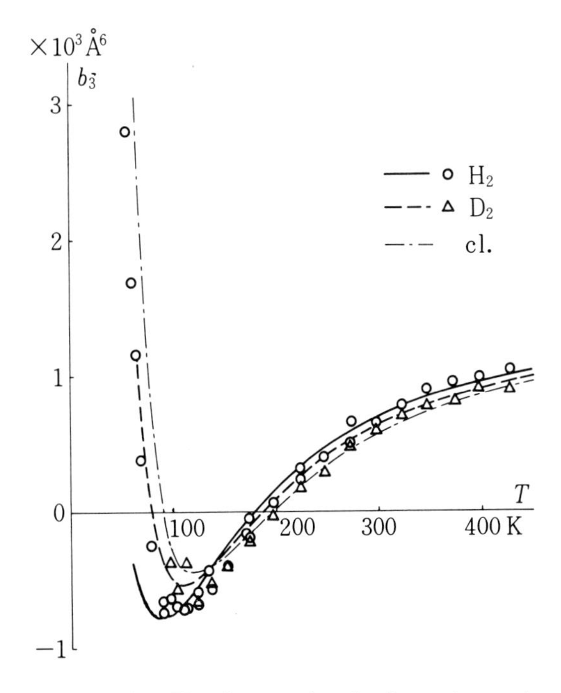

Figure 6.9 The three-molecule cluster integral for hydrogen and deuterium

the third term is

$$(b_3^{(1)})_{\nu=0} = \left(\frac{\lambda}{kT}\right)^{1/3} \sum_{t=0}^{\infty} I_t y^t. \tag{6.26}$$

The numerical values for the coefficients I_t are included in Table 6.2. For helium and hydrogen with large quantum effects, the nonadditive portion of the potential, that is, the second term on the right-hand side of (6.25), has little contribution.

In Figures 6.8 and 6.9, the computed values of b_3 with the aid of the potential already determined are compared with the measured values. The results may be regarded as satisfactory in the sense that the agreement between the computed values and the measured values is improved by taking account of quantum effects.

6.5 Potential between unlike rare-gas atoms

Thus far we have treated rare-gas atoms of the same kind; in this section we consider potential between rare-gas atoms of different kinds. The problem is the following. When the potentials.

$$\lambda_{AA} r^{-12} - \mu_{AA} r^{-6} \quad \text{and} \quad \lambda_{BB} r^{-12} - \mu_{BB} r^{-6}$$

between atoms AA and between atoms BB are known, how do we estimate the constants, λ_{AB} and μ_{AB}, in the potential between AB,

$$\lambda_{AB} r^{-12} - \mu_{AB} r^{-6}, \tag{6.27}$$

from λ_{AA}, λ_{BB}, μ_{AA}, μ_{BB}; and how is such an estimate compared with the measured results?

First we consider only the potential λr^{-12} representing the repulsive force. The repulsive force between two atoms AA at a distance $2r_1$ is $12\lambda_{AA}(2r_1)^{-13}$. The physical origin of the repulsive force may be interpreted as an 'elastic force' resulting from a compressed electron cloud; it may be said that when the atom A is compressed to a distance r_1 from the centre, a repulsive force of $12\lambda_{AA}(2r_1)^{-13}$ appears. Such a situation may hold to a good approximation when the other atom is not an atom A of the same kind. The repulsive force between two atoms AB at a distance r is $12\lambda_{AA}(2r_1)^{-13}$ and at the same time $12\lambda_{BB}(2r_2)^{13}$, where r_1 and r_2 are the distances from the 'boundary' to the centres of A and B, respectively. From

$$r_1 : r_2 = \lambda_{AA}^{1/13} : \lambda_{BB}^{1/13}$$

and $r = r_1 + r_2$, the repulsive force is

$$12\left(\frac{\lambda_{AA}^{1/13} + \lambda_{BB}^{1/13}}{2r}\right)^{13}.$$

Consequently the coefficient of the repulsive potential $\lambda_{AB} r^{-12}$ is determined as

$$\lambda_{AB}^{1/13} = \tfrac{1}{2}(\lambda_{AA}^{1/13} + \lambda_{BB}^{1/13}). \tag{6.28}$$

(This idea was first advanced by F. T. Smith (1972), *Phys. Rev.*, A **5**, 1708 and Chang Lyoul Kong (1973), *J. Chem. Phys.*, **59**, 2464.)

For the coefficient of the attractive potential $-\mu_{AB}r^{-6}$, it may be most natural to take

$$\mu_{AB} = (\mu_{AA}\mu_{BB})^{1/2} \tag{6.29}$$

on the basis of (4.23).

Writing the potential between AB in the form

$$U_{AB}(r) = U_{0AB}\left[\left(\frac{r_{0AB}}{r}\right)^{12} - 2\left(\frac{r_{0AB}}{r}\right)^{6}\right], \tag{6.30}$$

we may calculate r_{0AB} and U_{0AB} with the aid of the numerical values listed in Table 6.1; the results are shown in Table 6.3.

The second virial coefficient for a two-component mixed gas may be written in the form,

$$B_A x_A^2 + 2B_{AB}x_A x_B + B_B x_B^2,$$

where x_A and $x_B \equiv 1 - x_A$ are molar fractions of the component A and the component B; B_A and B_B are equal to the second virial coefficient of the pure gases, A and B. The 'mixed virial coefficient' B_{AB} is then related to $U_{AB}(r)$ via

$$B_{AB}(T) = 2\pi \int_0^\infty \left[1 - \exp\left(\frac{-U_{AB}(r)}{kT}\right)\right] r^2 dr.$$

Taking account of quantum corrections to this relationship, we may determine the potential coefficients with the aid of the measured values; the results,

He–Ne	$r_0 = 3.00$ Å	$U_0 = 19.0$ K
He–Ar	3.50	26.3
Ne–Ar	3.50	55.0

agree almost completely with the values in Table 6.3.

Table 6.3 The potential constants between rare-gas atoms of different kinds

	r_{0AB}(Å)	U_{0AB}/k (K)
He–Ne	3.02	18.6
He–Ar	3.52	25.6
He–Kr	3.74	26.1
He–Xe	4.09	24.3
Ne–Ar	3.52	57.1
Ne–Kr	3.72	60.9
Ne–Xe	4.02	59.9
Ar–Kr	3.99	139
Ar–Xe	4.23	153
Kr–Xe	4.36	190

When $r_{0\,\mathrm{AA}}$ and $r_{0\,\mathrm{BB}}$ are close to each other (for example, Ar–Kr), we may use

$$2r_{0\,\mathrm{AB}} = r_{0\,\mathrm{AA}} + r_{0\,\mathrm{BB}} \quad \text{and} \quad U_{0\,\mathrm{AB}}{}^2 = U_{0\,\mathrm{AA}} U_{0\,\mathrm{BB}} \tag{6.31}$$

as a good approximation to (6.28) and (6.29) even if $U_{0\,\mathrm{AA}}$ and $U_{0\,\mathrm{BB}}$ are substantially different. We shall recall this simple connecting formula in Section 7.4.

Chapter 7

Intermolecular Potential with Convex Cores

7.1 Core potential and fundamental measures of a convex body

The potentials $U(r)$ between rare-gas atoms are functions of the internuclear distance r only; the functional forms are mutually quite similar. In contrast, the potentials between polyatomic molecules are complex; their shapes and symmetries are in general different, depending upon the molecules. The core potential (Kihara 1953) makes it possible to take a certain account of individuality of the molecules and to carry out a systematic treatment of intermolecular potentials.

We consider an appropriate core inside each molecule. For example, we may choose as the cores the line segment connecting the two O atoms in carbon dioxide CO_2, a thin hexagonal plate formed by the six C atoms in benzene C_6H_6, and the regular tetrahedron constructed by the four F atoms in carbon tetrafluoride CF_4. We assume that the core does not deform and is a convex body, defined in the following way.

When a line segment connecting two arbitrarily chosen points inside a body is always contained in the body, it is called a *convex body*. Since an egg is such a typical example, a convex body is also called an ovaloid. A sphere, an ellipsoid, and a convex polyhedron of course represent convex bodies. In addition there are convex bodies with zero volume like a thin circular plate; a line segment (a thin rod) may be regarded as a convex body with zero volume and zero surface area.

Now let us introduce a model in which the potential between two molecules is a function $U(\rho)$ of only the shortest distance ρ between their cores. In such a model the molecular shape may be reasonably taken into account. A simplification is involved, however, in that the depth of the potential is assumed to be independent of the molecular orientation. How the potential depth actually depends on the molecular orientation differs entirely from molecule to molecule; the situation will be clarified in the following two chapters.

As the form of $U(\rho)$, it may be natural to adopt the Lennard-Jones function,

$$U(\rho) = U_0\left[\left(\frac{\rho_0}{\rho}\right)^{12} - 2\left(\frac{\rho_0}{\rho}\right)^{6}\right]. \tag{7.1}$$

In these circumstances the Lennard-Jones potential widely used for rare gases

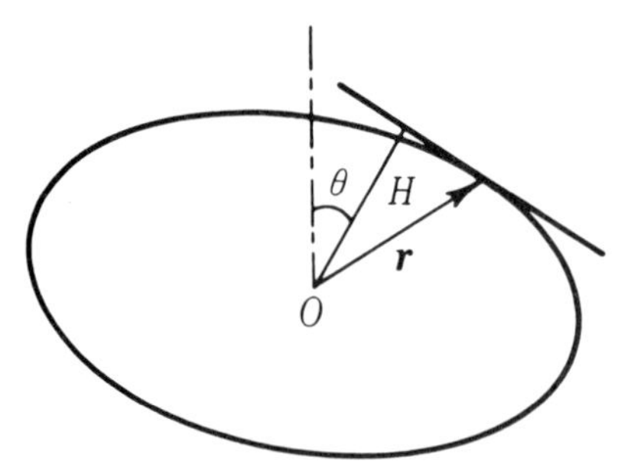

Figure 7.1 Supporting plane of a convex body, supporting function H, and the positon **r** of the supporting point

corresponds to a special case of the core potential in that the core reduces to a single point.

A special feature of this model lies in the fact that the expression for the second virial coefficient can be obtained as a theorem in the geometry of convex bodies. In other words, the statistics of molecular interaction reduce to geometrical properties of convex bodies representing the molecular cores.

Consider first a single convex body. Its volume V is a fundamental quantity representing a three-dimensional size; its surface area S is a fundamental quantity representing a two-dimensional size. As a fundamental quantity representing a one-dimensional size, we define the 'measure' M in the following way.

Inside a convex body we choose the origin O and the coordinate axes, and define the angular components, θ and φ $(0 \leqslant \theta \leqslant \pi, 0 \leqslant \varphi \leqslant 2\pi)$, of the polar coordinates in the usual way. For an arbitrary direction (θ, φ), there is only one plane that has a normal direction θ and φ from the origin and that is tangential to the convex body. Such a plane is called the *supporting plane* in the direction (θ, φ). Regarding the normal distance from the origin to the supporting plane as a function of θ and φ, we call it the *supporting function.* Usually the supporting function is denoted by $H(\theta, \varphi)$ (see Figure 7.1). Let M be the integral of the supporting function over all the directions:

$$M \equiv \int_0^{2\pi} \int_0^{\pi} H(\theta, \varphi) \sin\theta d\theta d\varphi. \tag{7.2}$$

It is clear that M is independent of the choice of the origin. When the convex body is a sphere with radius a, $M = 4\pi a$; hence in general $M/4\pi$ corresponds to an average radius of the convex body. In particular, for a thin rod with length l, such an average radius is a quarter of l. For, choosing the origin at the centre of the rod and the polar axis along the rod, we have

$$M = 2\pi \int_0^{\pi} \frac{l}{2} |\cos\theta| \sin\theta d\theta = \pi l.$$

The volume V, the surface area S, and the measure M are the *three fundamental measures* of a convex body. In the following section we study a convenient way of calculating M and summarize the three fundamental measures for a variety of typical convex bodies.

7.2 Steiner's formula

Let us assume, for the moment, that the convex body has a smooth surface and that its supporting plane makes a first-order contact with the convex body. Let $\mathbf{r}(\theta, \varphi)$ be the position vector of the point of contact (supporting point) between the convex body and the supporting plane in the direction (θ, φ) (Figure 7.1). With the aid of a unit vector $\mathbf{e}(\theta, \varphi)$ in the (θ, φ) direction, the supporting function can be expressed as

$$H(\theta, \varphi) = \mathbf{r}(\theta, \varphi) \cdot \mathbf{e}(\theta, \varphi). \tag{7.3}$$

Hence one of the fundamental measures, M, is given by

$$M = \iint \mathbf{r} \cdot \mathbf{e} \sin \theta d\theta d\varphi = \iint \mathbf{r} \cdot \left(\frac{\partial \mathbf{e}}{\partial \theta} \times \frac{\partial \mathbf{e}}{\partial \varphi}\right) d\theta d\varphi, \tag{7.4}$$

where we have used $(\partial \mathbf{e}/\partial \theta) \times (\partial \mathbf{e}/\sin \theta \partial \varphi) = \mathbf{e}$.

The surface area S of a convex body is given by

$$S = \iint \left| \frac{\partial \mathbf{r}}{\partial \theta} \times \frac{\partial \mathbf{r}}{\partial \varphi} \right| d\theta d\varphi.$$

Since $(\partial \mathbf{r}/\partial \theta) \times (\partial \mathbf{r}/\partial \varphi)$ is parallel to $\mathbf{e}$, we have

$$S = \iint \mathbf{e} \cdot \left(\frac{\partial \mathbf{r}}{\partial \theta} \times \frac{\partial \mathbf{r}}{\partial \varphi}\right) d\theta d\varphi. \tag{7.5}$$

The volume V of a convex body is obviously

$$V = \frac{1}{3} \iint \mathbf{r} \cdot \left(\frac{\partial \mathbf{r}}{\partial \theta} \times \frac{\partial \mathbf{r}}{\partial \varphi}\right) d\theta d\varphi. \tag{7.6}$$

For a sphere with a unit radius one obtains

$$4\pi = \iint \mathbf{e} \cdot \left(\frac{\partial \mathbf{e}}{\partial \theta} \times \frac{\partial \mathbf{e}}{\partial \varphi}\right) d\theta d\varphi \tag{7.7}$$

from (7.5) or (7.6). This constant is a quantity which may be called a zero-dimensional fundamental measure; it is used together with V, S, and M.

Now, for an arbitrary convex body A, its fundamental measures and $\mathbf{r}$ function are

$$V_{\mathrm{A}}, \quad S_{\mathrm{A}}, \quad M_{\mathrm{A}}, \quad \mathbf{r}_{\mathrm{A}}(\theta, \varphi).$$

We consider a convex body constructed by all those points whose shortest distances from the convex body A do not exceed ρ; such a convex body may be called a '*parallel body* of A with thickness ρ'. We denote such a parallel body by $\mathrm{A} + \rho$.

The $\mathbf{r}$ function for the parallel body $A + \rho$ is equal to $\mathbf{r}_A(\theta, \varphi) + \rho\mathbf{e}(\theta, \varphi)$. Furthermore, its fundamental measures are calculated with the aid of the foregoing integral representations in the following away:

$$M_{A+\rho} = \iint (\mathbf{r}_A + \rho\mathbf{e}) \cdot \left(\frac{\partial \mathbf{e}}{\partial \theta} \times \frac{\partial \mathbf{e}}{\partial \varphi}\right) d\theta d\varphi = M_A + 4\pi\rho, \tag{7.8}$$

$$S_{A+\rho} = \iint \mathbf{e} \cdot \left[\frac{\partial(\mathbf{r}_A + \rho\mathbf{e})}{\partial \theta} \times \frac{\partial(\mathbf{r}_A + \rho\mathbf{e})}{\partial \varphi}\right] d\theta d\varphi.$$

Integrating both sides of the identity

$$\frac{\partial}{\partial \theta}\left[\mathbf{e} \cdot \left(\mathbf{r}_A \times \frac{\partial \mathbf{e}}{\partial \varphi}\right)\right] + \frac{\partial}{\partial \varphi}\left[\mathbf{e} \cdot \left(\frac{\partial \mathbf{e}}{\partial \theta} \times \mathbf{r}_A\right)\right]$$
$$= \mathbf{e} \cdot \left(\frac{\partial \mathbf{r}_A}{\partial \theta} \times \frac{\partial \mathbf{e}}{\partial \varphi}\right) + \mathbf{e} \cdot \left(\frac{\partial \mathbf{e}}{\partial \theta} \times \frac{\partial \mathbf{r}_A}{\partial \varphi}\right) - 2\mathbf{r}_A \cdot \left(\frac{\partial \mathbf{e}}{\partial \theta} \times \frac{\partial \mathbf{e}}{\partial \varphi}\right)$$

with respect to θ and φ, we find that the integral on the left-hand side vanishes; we thus have

$$\iint \left[\mathbf{e} \cdot \left(\frac{\partial \mathbf{r}_A}{\partial \theta} \times \frac{\partial \mathbf{e}}{\partial \varphi}\right) + \mathbf{e} \cdot \left(\frac{\partial \mathbf{e}}{\partial \theta} \times \frac{\partial \mathbf{r}_A}{\partial \varphi}\right)\right] d\theta d\varphi = 2M_A.$$

With the aid of this relationship we find

$$S_{A+\rho} = S_A + 2M_A\rho + 4\pi\rho^2. \tag{7.9}$$

Furthermore,

$$V_{A+\rho} = V_A + \int_0^\rho S_{A+\rho} d\rho = V_A + S_A\rho + M_A\rho^2 + (4\pi/3)\rho^3. \tag{7.10}$$

Equations 7.8, 7.9 and 7.10 are called Steiner's formulas.

On the other hand, the surface area of the parallel body $A + \rho$ is equal to the integration of $(1 + \rho/R_1)(1 + \rho/R_2)$ over the surface of the convex body A, where R_1 and R_2 are the principal radii of curvature (Figure 7.2). Hence,

$$S_{A+\rho} = S_A + \rho \int \left(\frac{1}{R_1} + \frac{1}{R_2}\right) dS_A + \rho^2 \int \frac{dS_A}{R_1 R_2}.$$

Comparison of this with (7.9) yields in general

$$M = \int \frac{1}{2}\left(\frac{1}{R_1} + \frac{1}{R_2}\right) dS. \tag{7.11}$$

The measure M of a convex body is thus equal to integration of the mean curvature over the surface of the convex body.

The foregoing relations have been derived under the assumption that the convex body has a smooth surface and that the supporting plane makes a first-order contact with it. Since any convex body can be approximated by such an ideal convex body with as much accuracy as one wishes, however, the results are applicable to all convex bodies.

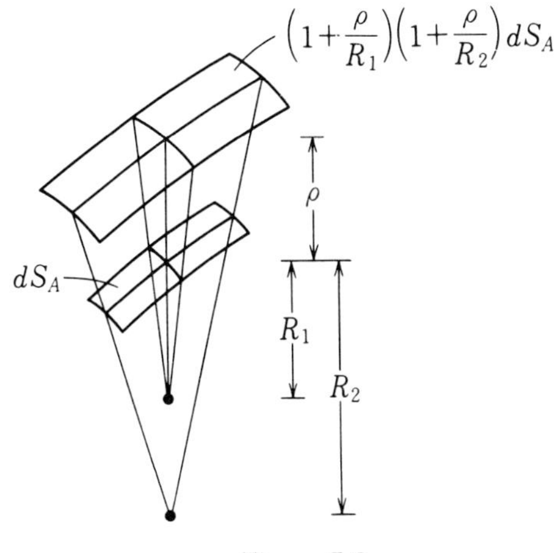

Figure 7.2

For the purpose of calculating M explicitly, (7.11) is convenient. If the convex body has edges, one may first consider its parallel body with thickness ρ and then take the limit $\rho \to 0$.

Problem

Show for a convex polyhedron

$$M = \frac{1}{2} \sum_{i=1}^{N} (\pi - \alpha_i) l_i,$$

where l_i and α_i are the length and the inner angle of the ith edge; N is the number of edges.

Solution

Let us denote the convex body by A. On the surface of the parallel body A + ρ constructed with thickness ρ out of A, there are as many spherical surfaces with radius ρ as there are vertices; as many cylindrical surfaces with radius ρ as there are edges; and as many plane portions as there are sides. Those constitute the surface of the parallel body. With the aid of (7.11), we have

$$M_{A+\rho} = \frac{1}{\rho} \times \text{(summation of the areas of the spherical portions)}$$
$$+ \frac{1}{2\rho} \times \text{(summation of the areas of the cylindrical portions).}$$

The summation of the areas of the spherical portions is $4\pi\rho^2$; the summation of the areas of the cylindrical portions is $\Sigma(\pi - \alpha_i)\rho l_i$. Taking the limit $\rho \to 0$, we obtain M_A.

Table 7.1 Fundamental measures of convex bodies adopted for molecular cores

---

Sphere with radius a

$$V = 4\pi a^3/3, \quad S = 4\pi a^2, \quad M = 4\pi a.$$

Regular tetrahedron inscribed in a unit sphere

$$V = 8\sqrt{3}/27, \quad S = 8\sqrt{3}/3, \quad M = 2\sqrt{6}\cos^{-1}(-1/3) = 9.360.$$

Octahedron inscribed in a unit sphere

$$V = 4/3, \quad S = 4\sqrt{3}, \quad M = 6\sqrt{2}\cos^{-1}(1/3) = 10.445.$$

Rectangular parallellepiped with circumference c and area a for the bottom side, and with height l

$$V = lf, \quad S = 2f + cl, \quad M = \pi l + \pi c/2.$$

Thin plate (in the limit $l \to 0$ for the case above)

$$V = 0, \quad S = 2f, \quad M = \pi c/2.$$

Thin rod with length l

$$V = 0, \quad S = 0, \quad M = \pi l.$$

Truncated double cone as shown in Figure 7.3

$$V = \frac{2\pi}{3}(1 - \lambda^3)R^3 \tan\alpha, \quad S = 2\pi\left(\lambda^2 + \frac{1 - \lambda^2}{\cos\alpha}\right)R^2,$$

$$M = \pi^2 R + 2\pi(1 - \lambda)(\tan\alpha - \alpha)R.$$

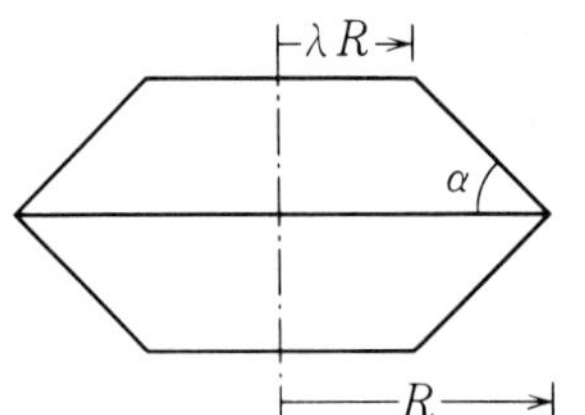

Figure 7.3 Truncated double cone. Core of a cyclohexane molecule

In Table 7.1 we summarize the fundamental measures of various convex bodies adopted for the cores of polyatomic molecules.

Among the fundamental measures of the convex body, a number of inequalities are known, such as Minkowski's inequality

$$S \leqslant (4\pi)^{-1} M^2. \tag{7.12}$$

When the convex body is a sphere, the equality in (7.12) holds; when the convex body resembles a sphere, both sides are quite close to each other.

7.3 Isihara–Hadwiger's formula

Consider two convex bodies A and B. For the moment we assume that both convex bodies have smooth surfaces and that any supporting plane makes a first-order contact with the convex bodies. The origins, O_A and O_B, of the coordinates are fixed inside those convex bodies; the three fundamental measures and the $\mathbf{r}$ functions for them are written as

$$V_A, \quad S_A, \quad M_A, \quad \mathbf{r}_A(\theta, \varphi),$$

$$V_B, \quad S_B, \quad M_B, \quad \mathbf{r}_B(\theta, \varphi).$$

Fixing O_A and the direction of A, we let B move with its direction fixed and keeping contact with A from outside. The trace of O_B then forms a convex body; this is denoted by A + B. Its $\mathbf{r}$ function is given by

$$\mathbf{r}(\theta, \varphi) = \mathbf{r}_A(\theta, \varphi) + \mathbf{r}_B^*(\theta, \varphi).$$

As shown in Figure 7.4, one has

$$\mathbf{r}_B^*(\theta, \varphi) = -\mathbf{r}_B(\pi - \theta, \pi + \varphi).$$

We must first of all calculate the volume of the convex body A + B,

$$V_{A+B} = \frac{1}{3} \iint (\mathbf{r}_A + \mathbf{r}_B^*) \cdot \left[\left(\frac{\partial \mathbf{r}_A}{\partial \theta} + \frac{\partial \mathbf{r}_B^*}{\partial \theta}\right) \times \left(\frac{\partial \mathbf{r}_A}{\partial \varphi} + \frac{\partial \mathbf{r}_B^*}{\partial \varphi}\right)\right] d\theta d\varphi.$$

For this purpose we note the identity

$$\frac{\partial}{\partial \theta}\left[\mathbf{r}_B^* \cdot \left(\mathbf{r}_A \times \frac{\partial \mathbf{r}_A}{\partial \varphi}\right)\right] + \frac{\partial}{\partial \varphi}\left[\mathbf{r}_B^* \cdot \left(\frac{\partial \mathbf{r}_A}{\partial \theta} \times \mathbf{r}_A\right)\right]$$

$$= 2\mathbf{r}_B^* \cdot \left(\frac{\partial \mathbf{r}_A}{\partial \theta} \times \frac{\partial \mathbf{r}_A}{\partial \varphi}\right) - \mathbf{r}_A \cdot \left(\frac{\partial \mathbf{r}_B^*}{\partial \theta} \times \frac{\partial \mathbf{r}_A}{\partial \varphi}\right) - \mathbf{r}_A \cdot \left(\frac{\partial \mathbf{r}_A}{\partial \theta} \times \frac{\partial \mathbf{r}_B^*}{\partial \varphi}\right).$$

Integrating both sides with respect to θ and φ, we find the integral on the left-hand side vanishes; with the aid of this we obtain

$$\iint \left[\mathbf{r}_A \cdot \left(\frac{\partial \mathbf{r}_B^*}{\partial \theta} \times \frac{\partial \mathbf{r}_A}{\partial \varphi}\right) + \mathbf{r}_A \cdot \left(\frac{\partial \mathbf{r}_A}{\partial \theta} \times \frac{\partial \mathbf{r}_B^*}{\partial \varphi}\right)\right] d\theta d\varphi$$

$$= 2 \iint \mathbf{r}_B^* \cdot \left(\frac{\partial \mathbf{r}_A}{\partial \theta} \times \frac{\partial \mathbf{r}_A}{\partial \varphi}\right) d\theta d\varphi.$$

Using this and another analogous relation, we calculate

$$V_{A+B} = V_A + V_B + \iint \mathbf{r}_B^* \cdot \left(\frac{\partial \mathbf{r}_A}{\partial \theta} \times \frac{\partial \mathbf{r}_A}{\partial \varphi}\right) d\theta d\varphi$$

$$+ \iint \mathbf{r}_A \cdot \left(\frac{\partial \mathbf{r}_B^*}{\partial \theta} \times \frac{\partial \mathbf{r}_B^*}{\partial \varphi}\right) d\theta d\varphi.$$

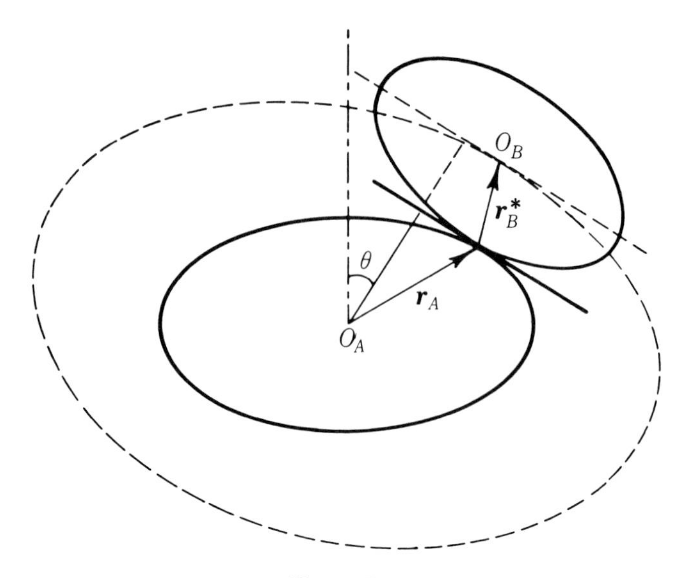

Figure 7.4

Thus far the direction of the convex body B has been kept unchanged. In the next step we carry out an average with respect to the directions of B. For this purpose we rewrite the integrand in the third term on the right-hand side of the foregoing formula as

$$\mathbf{r}_{\mathrm{B}}^* \cdot \mathbf{ee} \cdot \left(\frac{\partial \mathbf{r}_{\mathrm{A}}}{\partial \theta} \times \frac{\partial \mathbf{r}_{\mathrm{A}}}{\partial \varphi}\right).$$

Since the average of $\mathbf{r}_{\mathrm{B}}^* \cdot \mathbf{e}$ over the directions of the convex body B yields $(4\pi)^{-1}M_{\mathrm{B}}$, the average of the third term on the right-hand side is equal to $(4\pi)^{-1}M_{\mathrm{B}}S_{\mathrm{A}}$. Thus the average $\langle V_{\mathrm{A+B}}\rangle$ of $V_{\mathrm{A+B}}$ is expressed in terms of the fundamental measures of the two convex bodies in the following way:

$$\langle V_{\mathrm{A+B}}\rangle = V_{\mathrm{A}} + V_{\mathrm{B}} + (4\pi)^{-1}(M_{\mathrm{B}}S_{\mathrm{A}} + M_{\mathrm{A}}S_{\mathrm{B}}). \tag{7.13}$$

It is clear that this relationship applies for any convex bodies irrespective of the assumptions introduced in the beginning.

The formula (7.13) was established in 1950 by Isihara and by Hadwiger, independently (A. Isihara (1950), *J. Chem. Phys.*, **18**, 1446; H. Hadwiger (1950), *Mh. Math.*, **54**, 345). The elementary proof described above was due to the author (T. Kihara (1953), *Rev. Mod. Phys.*, **25**, 831).

We now consider a situation in which the convex body B moves at a distance ρ apart from A, rather than the convex body B contacting with the convex body A. The trace of O_{B} forms a surface of a still greater convex body $\mathrm{A}+\rho+\mathrm{B}$ in these circumstances. The average of the volume of such a convex body over the directions of B is

$$\langle V_{\mathrm{A}+\rho+\mathrm{B}}\rangle = V_{\mathrm{A+B}} + V_{\mathrm{B}} + (4\pi)^{-1}(M_{\mathrm{B}}S_{\mathrm{A}+\rho} + M_{\mathrm{A}+\rho}S_{\mathrm{B}}).$$

Substitution of Steiner's formula yields

$$\begin{aligned}\langle V_{A+\rho+B}\rangle = V_A + V_B + (4\pi)^{-1}(M_B S_A + M_A S_B) \\ + [S_A + S_B + (2\pi)^{-1} M_A M_B]\rho \\ + (M_A + M_B)\rho^2 + (4\pi/3)\rho^3.\end{aligned} \tag{7.14}$$

The average of the surface area of this convex body is

$$\begin{aligned}\langle S_{A+\rho+B}\rangle &= \frac{d}{d\rho}\langle V_{A+\rho+B}\rangle \\ &= S_A + S_B + (2\pi)^{-1} M_A M_B + 2(M_A + M_B)\rho + 4\pi\rho^2.\end{aligned} \tag{7.15}$$

Equations 7.13, 7.14, and 7.15 will be used in the following section and thereafter.

7.4 The second virial coefficient for the core potential

Extension of the expression for the second virial coefficient $B(T)$ for the rare gases,

$$2B(T) = \int_0^\infty (1 - e^{-U(r)/kT}) 4\pi r^2\, dr, \tag{7.16}$$

to the cases of polyatomic molecules takes the form

$$2B(T) = \left\langle \int (1 - e^{-U/kT}) d\tau \right\rangle. \tag{7.17}$$

Here $d\tau$ represents the volume element which the 'centre' of a molecule occupies relative to the other; $\langle \ldots \rangle$ implies an average with respect to the orientations of the molecule.

For the core potential, U is a function $U(\rho)$ of only the shortest distances ρ between the cores. Hence, (7.17) can be expressed as

$$2B(T) = \int_0^\infty (1 - e^{-U(\rho)/kT}) \langle S_{\mathrm{core}+\rho+\mathrm{core}} \rangle d\rho + \langle V_{\mathrm{core}+\mathrm{core}} \rangle. \tag{7.18}$$

Here the quantities in $\langle \ldots \rangle$ are obtained from (7.13) and (7.15) as

$$\langle S_{\mathrm{core}+\rho+\mathrm{core}} \rangle = 2[S + (4\pi)^{-1} M^2] + 4M\rho + 4\pi\rho^2,$$

$$\langle V_{\mathrm{core}+\mathrm{core}} \rangle = 2[V + (4\pi)^{-1} MS];$$

V, S, and M are the three fundamental measures of the core.

When $U(\rho)$ takes a form similar to the Lennard-Jones potential (7.1),

$$U(\rho) = U_0 \left[\left(\frac{\rho_0}{\rho}\right)^{12} - 2\left(\frac{\rho_0}{\rho}\right)^{6} \right], \tag{7.19}$$

the integration may be carried out as in the previous chapter; the result is

$$B(T) = (2\pi/3)\rho_0^3 F_3(z) + M\rho_0^2 F_2(z) + [S + (4\pi)^{-1} M^2]\rho_0 F_1(z) + V + (4\pi)^{-1} MS. \tag{7.20}$$

Here $z = U_0/kT$, and $F_s(z)$ is expressed as follows:

$$F_s(z) = \int_0^\infty \left[1 - \exp\left(-\frac{z}{\xi^{12}} + 2\frac{z}{\xi^6}\right)\right] d(\xi^s)$$

$$= -\frac{s}{12} \sum_{t=0}^{\infty} \frac{1}{t!} \Gamma\left(\frac{6t-s}{12}\right) 2^t z^{(6t+s)/12}.$$

The numerical values of $F_3(z)$, $F_2(z)$, and $F_1(z)$ are shown in Table 7.2.

We have summarized the molecular structures in Section 2.1. Referring to the numerical values in Table 2.1, we choose the molecular cores as in Table 7.3. The

Table 7.2 Values of the three functions in equation (7.20)

$-\log_{10} z$	$F_3(z)$	$F_2(z)$	$F_1(z)$
−0.50	−17.115	−9.614	−3.833
−0.45	−12.834	−7.003	−2.614
−0.40	−9.859	−5.211	−1.784
−0.35	−7.721	−3.939	−1.199
−0.30	−6.138	−3.008	−0.7761
−0.25	−4.936	−2.311	−0.4613
−0.20	−4.003	−1.776	−0.2221
−0.15	−3.265	−1.358	−0.0368
−0.10	−2.673	−1.027	0.1092
−0.05	−2.191	−0.7599	0.2257
0.00	−1.795	−0.5424	0.3198
0.1	−1.189	−0.2150	0.4600
0.2	−0.7587	0.0132	0.5561
0.3	−0.4465	0.1758	0.6234
0.4	−0.2170	0.2930	0.6710
0.5	−0.0469	0.3779	0.7045
0.6	0.0794	0.4392	0.7279
0.7	0.1729	0.4830	0.7436
0.8	0.2415	0.5134	0.7536
0.9	0.2911	0.5336	0.7591
1.0	0.3259	0.5459	0.7611
1.1	0.3493	0.5521	0.7604
1.2	0.3638	0.5534	0.7574
1.3	0.3715	0.5510	0.7528
1.4	0.3737	0.5457	0.7468
1.5	0.3718	0.5381	0.7396
1.6	0.3668	0.5287	0.7320
1.7	0.3594	0.5179	0.7228
1.8	0.3501	0.5062	0.7135
1.9	0.3396	0.4937	0.7038
2.0	0.3281	0.4806	0.6937

$F_3(z)$ is equal to $F(z)$ in (6.6).

Table 7.3 Fundamental measures of molecular cores and potential constants

Molecule	Core	$V(\text{Å}^3)$	$S(\text{Å}^2)$	$M(\text{Å})$	$\rho_0(\text{Å})$	$U_0/k(\text{K})$
H_2	line segment connecting H × 0.65	0	0	1.51	2.96	41.5
N_2	line segment connecting N × 0.85	0	0	2.92	3.60	117
O_2	line segment connecting O × 0.85	0	0	3.23	3.20	151
F_2	line segment connecting F	0	0	4.52	2.90	178
CO_2	line segment connecting O	0	0	7.23	3.30	316
CS_2	line segment connecting S	0	0	9.67	3.60	680
C_2H_4	line segment connecting C	0	0	4.18	4.10	264
C_2H_6	cylinder, height 1.6 Å, radius 0.3 Å	0.45	3.58	7.98	3.10	422
cyclo-C_6H_{12}	truncated double cone*	14.7	32.9	21.1	2.70	955
$C(CH_3)_4$	sphere, radius 1.10 Å	5.58	15.20	13.82	4.00	588
$Si(CH_3)_4$	sphere, radius 1.27 Å	8.58	20.26	15.95	4.00	631
CH_4	regular tetrahedron connecting the middle points between C and H	0.084	1.38	5.12	3.15	223
CF_4	regular tetrahedron connecting F	1.29	8.54	12.73	2.63	352
SiF_4	regular tetrahedron connecting F	1.86	10.91	14.39	2.90	372
C_6H_6	hexagon connecting C	0	10.04	13.10	3.40	850
$(CH_2)_3$	triangle connecting C	0	2.01	7.18	2.70	716

*R = 2.0 Å, λ = 0.5, and $\alpha = \pi/4$ in Figure 7.3.

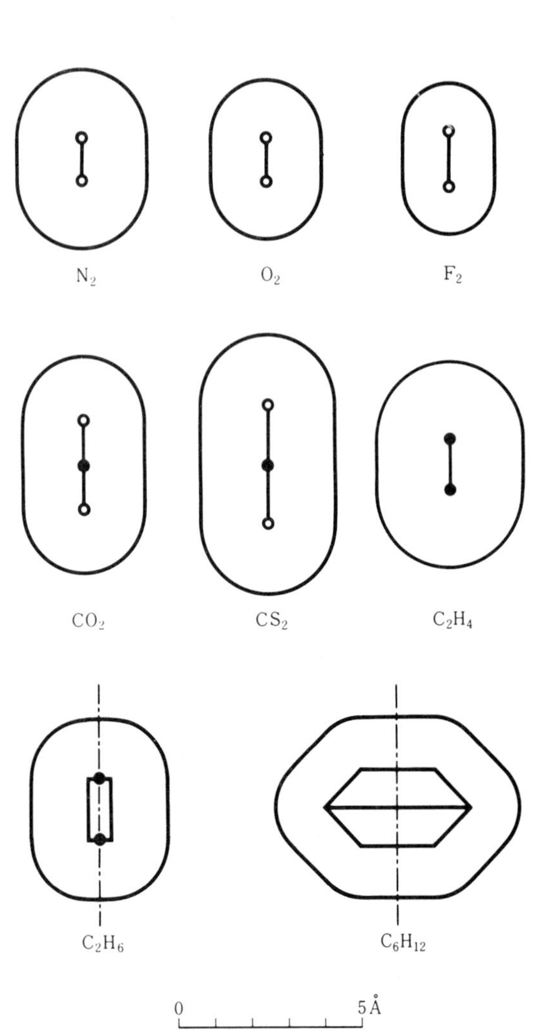

Figure 7.5 Shapes of molecules depicted on the basis

fundamental measures – volume V, surface area S, and the integral of mean curvature M – are then determined as listed in Table 7.3.

Substituting those fundamental measures in (7.20), we may determine the two potential constants, ρ_0 and U_0/k, in such a way that the measured values of the second virial coefficient agree with the calculated curves. The results are also listed in Table 7.3; for H_2 the quantum effects are taken into account.

Figure 7.5 depicts parallel bodies in which a thickness $\rho_0/2$ is added to a molecular core; it may be regarded as a way of representation for the shape and size of a molecule.

Basically the core potential is a model for a nondeformable molecule. It has

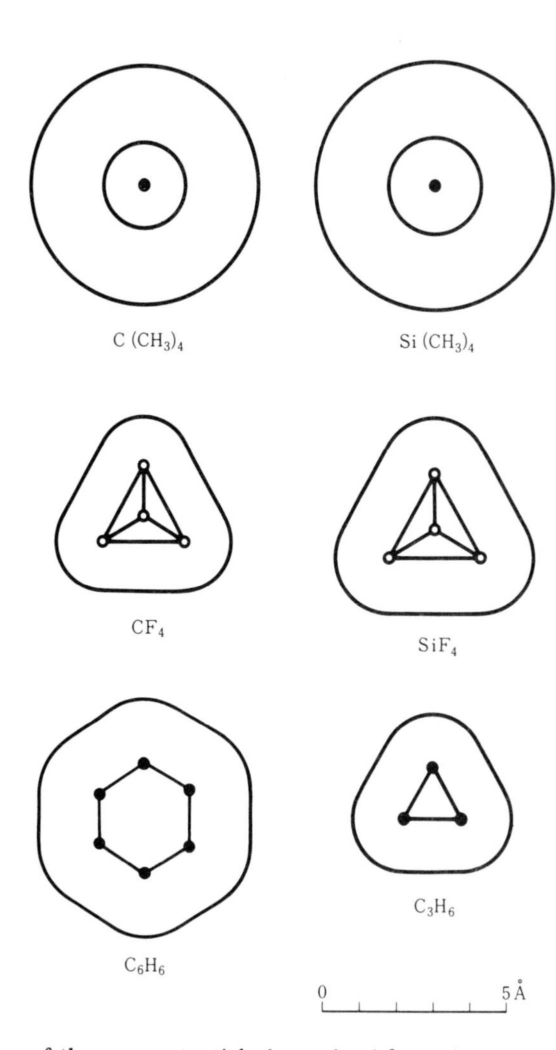

of the core potentials determined from the second virial coefficients

yielded unexpectedly successful results, nevertheless, in the application to such molecules as n-paraffin which do not have a fixed shape (Connolly and Kandalic 1960). The core adopted is a plate of rectangular shape as shown in Figure 7.6; comparison between measured values and the curves based on the parameters in Table 7.4 are also shown in the figure.

For conclusion of this section we describe expressions for mixed gases. The second virial coefficient for a two-component mixed gas may be written in the form

$$B_A x_A^2 + 2B_{AB} x_A x_B + B_B x_B^2,$$

Here x_A and $x_B \equiv 1 - x_A$ are the molar fractions of the components A and B; B_A

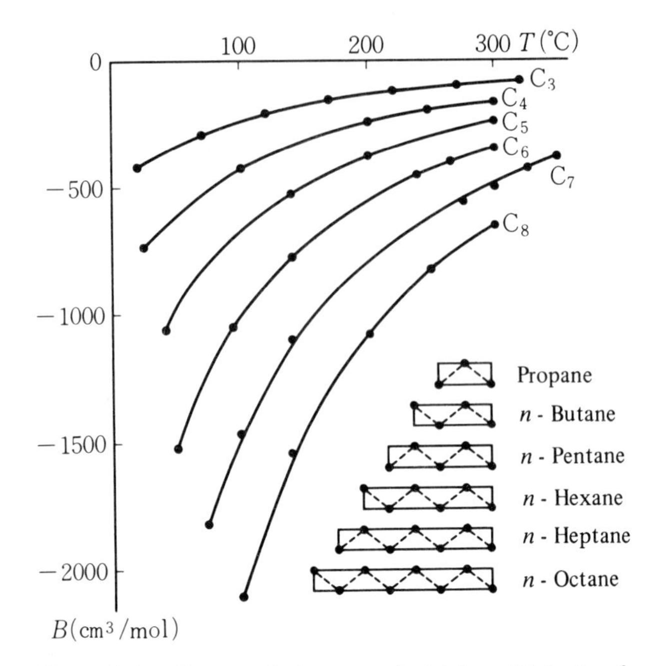

Figure 7.6 Cores and the second virial coefficients of *n*-paraffin. From J.F. Connolly and G.A. Kandalic (1960), *Phys. Fluids, 3,* 463

and B_B are the second virial coefficients for pure gases, A and B; and B_{AB}, called the mixed virial coefficient, is given by

$$2B_{AB} = \int_0^\infty \left[1 - \exp\left(\frac{-U_{AB}(\rho)}{kT}\right)\right] \langle S_{A+\rho+B}\rangle \, d\rho + \langle V_{A+B}\rangle . \qquad (7.21)$$

It has been assumed here that the potential $U_{AB}(\rho)$ between the AB molecules is a function of the shortest distance ρ between the two cores; $\langle V_{A+B}\rangle$ and $\langle S_{A+\rho+B}\rangle$ are given by (7.13) and (7.15) in terms of the fundamental measures of the two cores.

Table 7.4 Core potential for n-paraffin

	Core width (Å)	Core length (Å)	ρ_0(Å)	U_0/k(K)
propane	0.89	2 × 1.26	2.70	660
n-butane	0.89	3	2.70	795
n-pentane	0.89	4	2.70	890
n-hexane	0.89	5	2.70	985
n-heptane	0.89	6	2.70	1075
n-octane	0.89	7	2.70	1150

In particular, when the potential $U_{AB}(\rho)$ takes the form of the Lennard-Jones function,

$$U_{AB}(\rho) = U_{0AB}\left[\left(\frac{\rho_{0AB}}{\rho}\right)^{12} - 2\left(\frac{\rho_{0AB}}{\rho}\right)^{6}\right], \tag{7.22}$$

we have

$$\begin{aligned} 2B_{AB} = {} & (4\pi/3)\rho_{0AB}{}^3 F_3(z_{AB}) + (M_A + M_B)\rho_{0AB}{}^2 F_2(z_{AB}) \\ & + [S_A + S_B + (2\pi)^{-1} M_A M_B]\rho_{0AB} F_1(z_{AB}) \\ & + V_A + V_B + (4\pi)^{-1}(M_B S_A + M_A S_B), \end{aligned}$$

where $z_{AB} \equiv U_{0AB}/kT$. Here V_A, S_A, and M_A are the fundamental measures of the core A; V_B, S_B, and M_B are the fundamental measures of the core B; and the functions $F_s(z)$ are the same as those introduced earlier.

Based on detailed examination through comparison with measured values, one knows that

$$2\rho_{0AB} \equiv \rho_{0AA} + \rho_{0BB}, \quad U_{0AB}{}^2 = U_{0AA}U_{0BB} \tag{7.23}$$

represent good approximations for the potential constants. From the point of view of the core potential, we expect that ρ_0 does not directly depend on the sizes of molecules and may take on a similar magnitude; recalling the remark at the end of Section 6.5, we may also derive the foregoing connecting formula through theoretical means.

7.5 Three-convex-body problems

Regarding the potential between three molecules as a summation of potentials between two molecules, we may extend (5.44) and express the third virial coefficient $C(T)$ for polyatomic molecules as

$$3C(T) = \left\langle \int\int (1 - \exp(-U_{12}/kT))(1 - \exp(-U_{13}/kT))(1 - \exp(-U_{23}/kT)) d\tau_2 d\tau_3 \right\rangle$$

Here, $d\tau_2$ and $d\tau_3$ are the volume elements which the 'centres' of molecules 2 and 3 occupy when they move around a molecule 1 with their orientations fixed; $\langle \cdots \rangle$ means average with respect to the orientations of the molecules 2 and 3.

We again assume that the potential between two molecules is a function $U(\rho)$ of only the shortest distance ρ between those convex cores. For the shape of $U(\rho)$, let us first adopt the simplest case of a square-well type:

$$U(\rho) = \begin{cases} \infty & \rho < \sigma \\ -\epsilon < 0 & \sigma < \rho < 2\sigma \\ 0 & 2\sigma < \rho. \end{cases} \tag{7.24}$$

Here ϵ is a constant measuring the depth of the potential, and σ is a constant defined in such a way that a parallel body constructed with an added thickness $\sigma/2$

to the core represents the hard portion of the molecule. The coefficient 2 in the formula above has been chosen so that for rare gases, i.e. when the core is a point, it agrees with the case of $g = 2$ in (5.47). Expression (5.46) with $g < 2$ corresponds to (7.24) for a spherical core.

The second virial coefficient $B(T)$ can easily be calculated. The integral of (7.17) may be expressed in the following form:

$$2B(T) = J^{(0)} - xJ^{(1)},$$

where

$$x \equiv e^{\epsilon/kT} - 1, \tag{7.25}$$

and $J^{(0)}$ and $J^{(1)}$ represent integrals $\langle \int d\tau \rangle$ over domains $\rho < \sigma$ and $\sigma < \rho < 2\sigma$, respectively. With the aid of the symbols used in the previous two sections,

$$J^{(0)} = \langle V_{\text{core}+\sigma+\text{core}} \rangle,$$
$$J^{(1)} = \langle V_{\text{core}+\sigma+\text{core}} \rangle.$$

Expressing these in terms of the three fundamental measures, M, S, and V, of the core, we finally obtain

$$B(T) = (2\pi/3)\sigma^3(1 - 7x) + M\sigma^2(1 - 3x)$$
$$+ [S + (4\pi)^{-1}M^2]\,\sigma(1 - x) + V + (4\pi)^{-1}MS. \tag{7.26}$$

The third virial coefficient is written as a cubic formula of x:

$$3C(T) = I^{(0)} - 3xI^{(1)} + 3x^2I^{(2)} - x^3I^{(3)}, \tag{7.27}$$

where $I^{(i)}$ represent the integrals $\langle \int\int d\tau_2\, d\tau_3 \rangle$ over the following domains:

$$I^{(0)}:\ \rho_{12} < \sigma, \qquad \rho_{13} < \sigma, \qquad \rho_{23} < \sigma$$
$$I^{(1)}:\ \sigma < \rho_{12} < 2\sigma, \qquad \rho_{13} < \sigma, \qquad \rho_{23} < \sigma$$
$$I^{(2)}:\ \sigma < \rho_{12} < 2\sigma, \qquad \sigma < \rho_{13} < 2\sigma, \qquad \rho_{23} < \sigma$$
$$I^{(3)}:\ \sigma < \rho_{12} < 2\sigma, \qquad \sigma < \rho_{13} < 2\sigma, \qquad \sigma < \rho_{23} < 2\sigma.$$

ρ_{12} is the distance between the cores of molecules 1 and 2.

To evaluate those integrals, we define a function $F(\mathrm{A, B, C})$ of the three convex bodies A, B, and C as follows. Fixing the convex body A, we let the convex bodies B and C move, keeping their orientations unchanged, over those configurations in which the spaces between A and B, between A and C, and between B and C all overlap. In these circumstances, 'centres' of B and C altogether form a 6-dimensional domain. $F(\mathrm{A, B, C})$ is defined as an average of such a 6-dimensional volume over the orientations of B and C. With the aid of this function, $I^{(0)}, \ldots, I^{(3)}$ are given as follows:

$$I^{(0)} = F(\text{core} + \sigma/2, \text{core} + \sigma/2, \text{core} + \sigma/2),$$
$$I^{(1)} = F(\text{core} + \sigma, \text{core} + \sigma, \text{core}) - I^{(0)},$$
$$I^{(2)} = F(\text{core} + 3\sigma/2, \text{core} + \sigma/2, \text{core} + \sigma/2) - I^{(0)} - 2I(1),$$
$$I^{(3)} = F(\text{core} + \sigma, \text{core} + \sigma, \text{core} + \sigma) - I^{(0)} - 3I^{(1)} - 3I^{(2)}.$$

The function $F(\mathrm{A, B, C})$ is an extension of $\langle V_{\mathrm{A+B}} \rangle$ to three-convex-body systems.

For a special case, when one of the three, e.g. A, is far greater than the other two, $F(\mathrm{A, B, C})$ becomes $V_{\mathrm{A}} \langle V_{\mathrm{B+C}} \rangle$, that is, it takes the form,

$$V_{\mathrm{A}} [V_{\mathrm{B}} + V_{\mathrm{C}} + (4\pi)^{-1} (S_{\mathrm{B}} M_{\mathrm{C}} + S_{\mathrm{C}} M_{\mathrm{B}})] .$$

Taking account of this asymptotic form, we find that the function can be written as

$$\begin{aligned} F(\mathrm{A, B, C}) = {} & V_{\mathrm{A}} V_{\mathrm{B}} + V_{\mathrm{A}} V_{\mathrm{C}} + V_{\mathrm{B}} V_{\mathrm{C}} \\ & + (4\pi)^{-1} [V_{\mathrm{A}}(S_{\mathrm{B}} M_{\mathrm{C}} + S_{\mathrm{C}} M_{\mathrm{B}}) + V_{\mathrm{B}}(S_{\mathrm{A}} M_{\mathrm{C}} + S_{\mathrm{C}} M_{\mathrm{A}}) \\ & + V_{\mathrm{C}}(S_{\mathrm{A}} M_{\mathrm{B}} + S_{\mathrm{B}} M_{\mathrm{A}}) + G(\mathrm{A, B, C})] . \end{aligned} \tag{7.28}$$

The problem thus reduces to evaluation of the last term, $G(\mathrm{A, B, C})$.

When all of the three bodies are spheres with radii a, b, and c, the function F can be calculated with the aid of the function W introduced in Section 5.7:

$$\begin{aligned} F(\mathrm{A, B, C}) &= W(a + b, a + c, b + c) \\ &= (4\pi)^2 [3^{-2} (a^3 b^3 + a^3 c^3 + b^3 c^3) \\ &\quad + 3^{-1} abc(a^2 b + ab^2 + a^2 c + ac^2 + b^2 c + bc^2) + a^2 b^2 c^2] , \end{aligned}$$

or

$$G(\mathrm{A, B, C}) = (4\pi)^3 a^2 b^2 c^2 .$$

Among those single-term formulas (or monomials) constructed from the fundamental measures of the three convex bodies which become equal to $(4\pi)^3 a^2 b^2 c^2$ in the cases of spheres, the smallest one is $S_{\mathrm{A}} S_{\mathrm{B}} S_{\mathrm{C}}$ and the largest one is $(4\pi)^{-3} M_{\mathrm{A}}^2 M_{\mathrm{B}}^2 M_{\mathrm{C}}^2$. Consequently, G is expected to lie somewhere between those two values (T. Kihara and K. Miyoshi (1975), *J. Stat. Phys.*, **13**, 337):

$$S_{\mathrm{A}} S_{\mathrm{B}} S_{\mathrm{C}} \leqslant G(\mathrm{A, B, C}) \leqslant (4\pi)^{-3} M_{\mathrm{A}}^2 M_{\mathrm{B}}^2 M_{\mathrm{C}}^2 . \tag{7.29}$$

This inequality has not been proven rigorously. However, by means of electronic computers and other methods, it has been confirmed almost completely.

For the present purpose of application to molecules, it is possible to use the arithmetic mean between the upper and lower limits of this inequality as an approximate value of G.

As an example let us consider the CF_4 molecule. Taking as the core the regular tetrahedron connecting the four F atoms, we list its fundamental measures in Table 7.3. Comparison between the measured values of the second virial coefficient and the formula (7.26) yield the constants of the square-well potential as

$$\sigma = 2.28 \text{ Å}, \quad \epsilon/k = 157 \text{ K}.$$

Figure 7.7 compares the two-molecule and three-molecule cluster integrals with measured values; for comparison the case in which the core reduces to a point is also depicted.

When the core is sufficiently small, that is, when $(4\pi)^{-1} M^2$ is negligible as compared with σ^2, the expression (7.26) for the second virial coefficient is

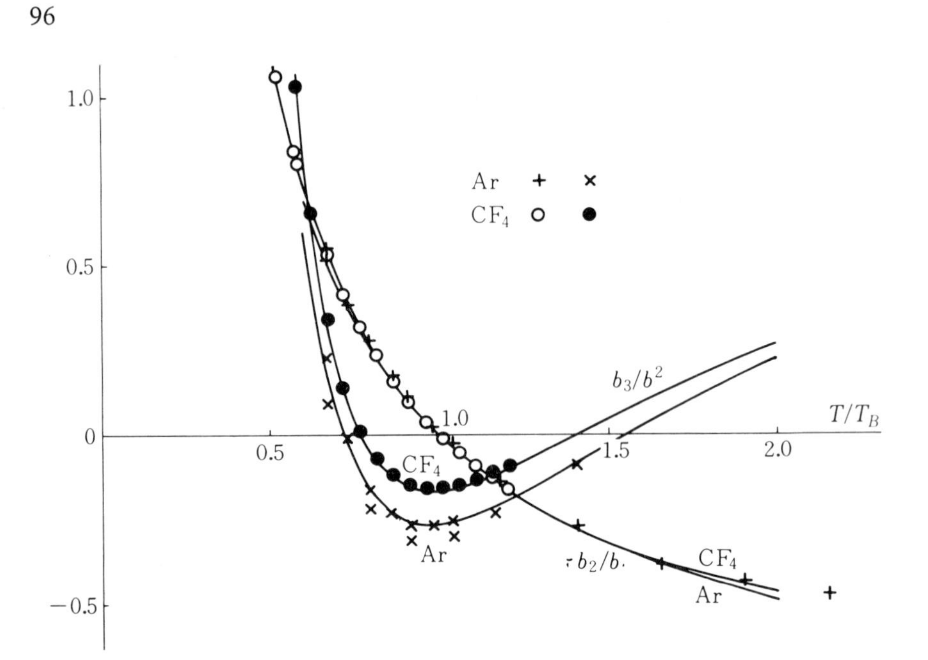

Figure 7.7 The two-molecule and three-molecule cluster integrals for square-well potentials with convex-body cores. The core of carbon tetrafluoride is the regular tetrahedron connecting F atoms; the core of argon is a point

simplified as

$$B(T) = (2\pi/3)\sigma^3(1-7x) + M\sigma^2(1-3x),$$

and the third virial coefficient is accurately calculated as

$$C(T) = (\pi^2/18)\sigma^6(5-17x+136x^2-162x^3)$$
$$+ (\pi/6)M\sigma^5(5-17x+84x^2-54x^3).$$

As may be clear from these, when the core is sufficiently small, the virial coefficients do not change even if one replaces the convex core by a sphere with an equal one-dimensional fundamental measure M, that is, a sphere with radius $M/4\pi$. (σ and ϵ should be kept unchanged.)

Even when the core is not small, for an arbitrary square-well potential (7.24), there is a square-well potential with a spherical core,

$$U(\rho) = \begin{cases} \infty & \rho < \sigma' \\ -\epsilon < 0 & \sigma' < \rho < 2\sigma' \\ 0 & 2\sigma' < \rho, \end{cases}$$

which gives the same second virial coefficient as (7.26). Here ρ is the shortest distance between the new spherical cores; ρ' is a new potential constant close to σ.

It is expected that the third virial coefficient calculated with regard to this square-well potential with a spherical core may represent a good approximation to $C(T)$ given by (7.27). That this is indeed the case has been actually confirmed.

7.6 Three-molecule cluster integrals for polyatomic molecules

It may be clear that the approximation scheme described in the last part of the previous section can be applied also to a more realistic core potential. Specifically we may state as follows.

The third virial coefficient with respect to the intermolecular potential (7.19) with a convex-body core,

$$U(\rho) = U_0 \left[\left(\frac{\rho_0}{\rho} \right)^{12} - 2 \left(\frac{\rho_0}{\rho} \right)^{6} \right],$$

can be calculated by means of a spherically symmetric core potential,

$$U(\rho) = U_0 \left[\left(\frac{\rho_0'}{\rho} \right)^{12} - 2 \left(\frac{\rho_0'}{\rho} \right)^{6} \right], \qquad (7.30)$$

which reproduces $B(T)$ with the same U_0. The radius a of the spherical core is then close to the $M/4\pi$ of the original convex-body core; ρ_0' is close to the original ρ_0. The values of those a and ρ_0' are listed in Table 7.5.

Table 7.5 Spherical approximation to the convex-body core

	Convex-body core		Spherical approximation	
	$(4\pi)^{-1} M$ (Å)	ρ_0 (Å)	α (Å)	ρ_0' (Å)
N_2	0.23	3.60	0.20	3.62
O_2	0.26	3.20	0.23	3.22
CO_2	0.58	3.30	0.46	3.40
CH_4	0.41	3.15	0.39	3.16
CF_4	1.01	2.63	0.90	2.70

For the cases of a spherical core, the numerical table necessary for the calculation of the third virial coefficient $C(T)$ is fortunately available (A. E. Sherwood and J. M. Prausnitz (1964), *J. Chem. Phys.*, **41**, 413). With the aid of this, the three-molecule cluster integral b_3 is computed from $b_3 = 2B^2 - C/2$; Figures 7.8–7.10 compare these with measured values. Additivity of the potential has been assumed here. Recalling that the effect of nonadditivity acts to shift the calculated curves of b_3 somewhat downward (Section 6.2), we may regard the results for O_2 in these figures as satisfactory. For N_2, it has become clear through detailed investigations that the effects of nonadditivity partially cancel with the effects of electric quadrupoles of the molecules. For CF_4, the disagreement may

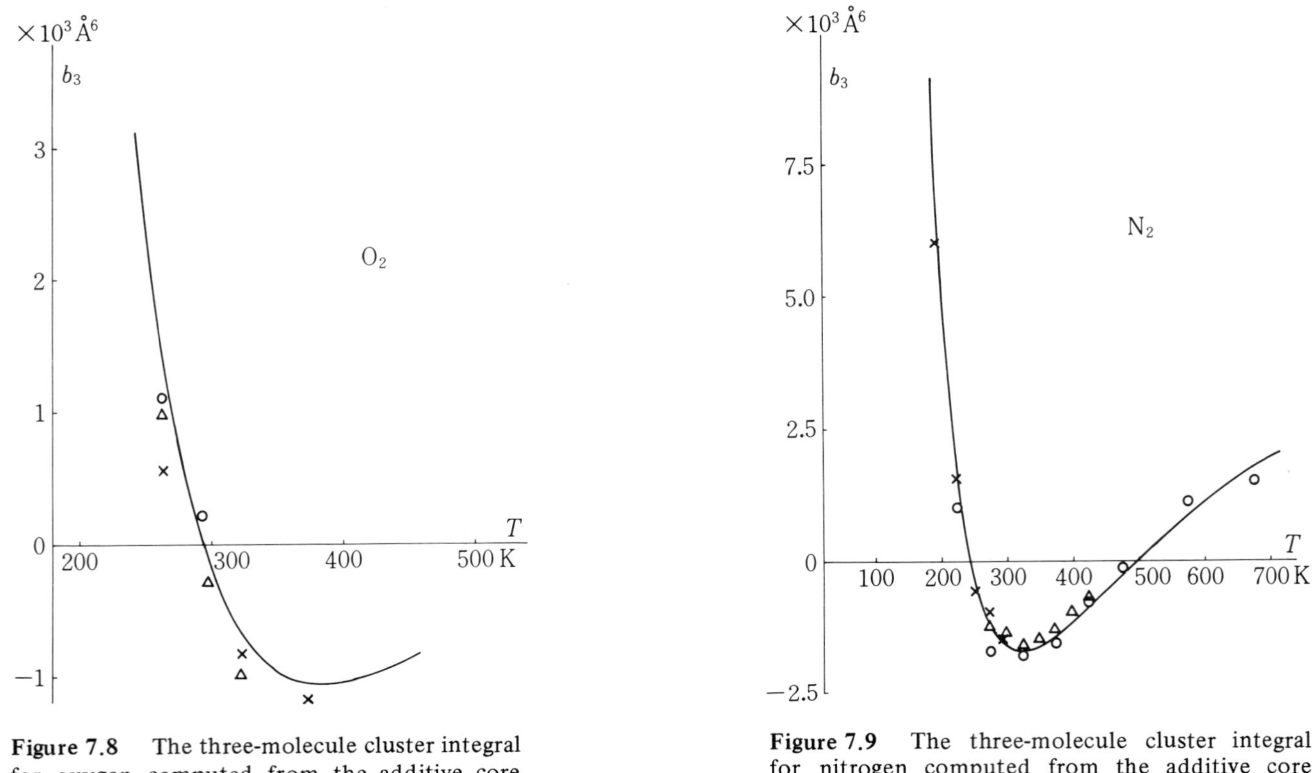

Figure 7.8 The three-molecule cluster integral for oxygen computed from the additive core potential, and the measured values

Figure 7.9 The three-molecule cluster integral for nitrogen computed from the additive core potential, and the measured values

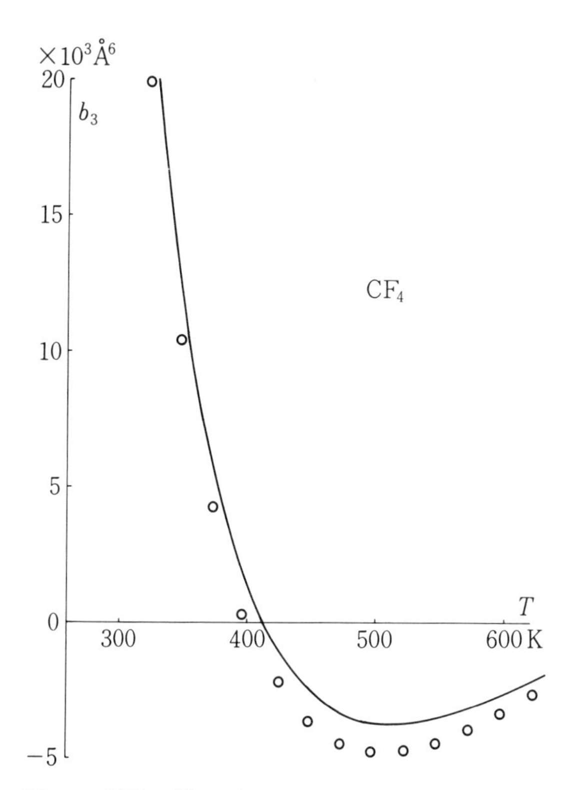

Figure 7.10 The three-molecule cluster integral for carbon tetrafluoride computed from the additive core potential, and the measured values

partially be due to the nonadditivity and partially due to the largeness of the cores. The effects of the core size have been described in Section 5.7. The regular tetrahedron connecting the four F atoms certainly tends to be a little too large for the core of CF_4.

Chapter 8

Potential Depth Depending on Molecular Orientations

8.1 The roles of electric multipoles

When molecules have electric multipoles, interactions between them are not so important in gases; they mostly vanish after averaging with respect to molecular orientations. In a solid state, on the other hand, the force between multipoles plays an important role.

Let us note the ratio between the temperature T_t at the triple point and the temperature T_c at the critical point. For each of Ne, Ar, Kr, and Xe, this ratio is 0.55; the law of corresponding states holds. In the following we list the values of T_t/T_c for symmetric linear molecules in the order of increasing magnitude:

carbon disulfide CS_2	0.29
oxygen O_2	0.35
fluorine F_2	0.37
chlorine Cl_2	0.41
hydrogen H_2	0.42
deuterium D_2	0.49
nitrogen N_2	0.50
cyanogen $(CN)_2$	0.61
acetylene C_2H_2	0.62
carbon dioxide CO_2	0.71.

Let us consider the meaning of this order. Since the molecular shapes are quite similar, certainly it does not represent the effects of the shape. In the vicinity of the critical point, the molecules while performing free rotation move around rather freely; potentials between molecules are averaged with respect to the directions of the molecules at almost equal probabilities. On the other hand, in a solid, that is in a molecular crystal, each molecule is arranged in its own characteristic configuration and thereby contributes to reduction of the total energy. Consequently, for those materials whose potential depths between molecules depend strongly on the molecular orientations, the temperature range of the solid phase is large as compared with that of the liquid phase, that is, T_t/T_c takes on a large value.

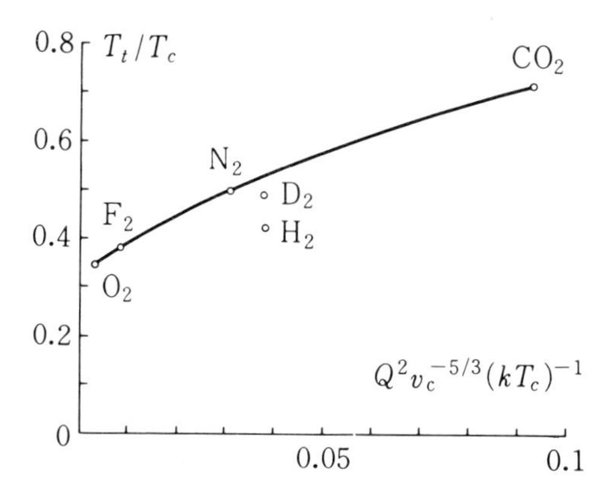

Figure 8.1 The ratio between the temperature T_t at the triple point and the temperature T_c at the critical point versus the reduced dimensionless strength of the quadrupole interactions

We now compute $Q^2 v_c^{-5/3} (kT_c)^{-1}$, a dimensionless quantity representing quadrupole interactions; Q is the quadrupole moment discussed in Section 3.3, and v_c is the volume per molecule at the critical point. The results are as follows:

O_2 0.002,
F_2 0.008,
N_2 0.027,
D_2 0.036,
H_2 0.038,
CO_2 0.097.

As we see in Figure 8.1, T_t/T_c is large for those molecules with large values of this quantity. When the molecular shapes are axisymmetric and approximately similar to each other, the solid phase is stable over a wide range of temperatures for those molecules which have relatively large quadrupole interactions.

Combining with foregoing statements, we see the following. In the case of linear symmetric molecules, the principal origin of the intermolecular-potential depth depending on the molecular orientations is the electrostatic force between the quadrupoles.

While cyclohexane C_6H_{12} has little quadrupole, benzene C_6H_6 has a substantial strength of quadrupole owing to the π electrons. The values of T_t/T_c, however, are 0.506 and 0.496, respectively; that is, they are almost equal. In the benzene crystal, it thus appears that the electrostatic energy between the quadrupoles does not considerably stabilize the actual crystal structure. This is related to the fact that the molecular configuration in the crystal is restricted due to the six H atoms extending around a molecule (cf. Section 9.4).

Carbon tetrafluoride CF_4 and silicon tetrafluoride SiF_4 are alike as far as the molecular shapes are concerned. Their values of the ratio T_t/T_c, however, are 0.39

Table 8.1 Effects of quadupolar force on solubilities

Solute	$\dfrac{c/c_0 \text{ in benzene}}{c/c_0 \text{ in cyclohexane}}$	Temperature
Ne	0.70	20 ~ 40 °C
Ar	0.70	20 ~ 40 °C
Kr	0.70	20 ~ 40 °C
Xe	0.68	20 ~ 40 °C
CH_4	0.77	25 °C
N_2	0.80	25 °C
CO_2	{1.58	20 °C
	{1.52	34 °C
C_2H_2	1.75	4 °C

and 0.76, respectively; they are remarkably different. This again can be understood from the fact that the octopole of SiF_4 is expected to be substantially stronger than the octopole of CF_4.

The solid carbon dioxide sublimates at –78.5 °C (at 1 atm); acetylene C_2H_2, at –84 °C; SiF_4, at –95 °C; SF_6, at –63 °C; UF_6, at 56.5 °C; and hexamethylenetetramine $(CH_2)_6N_4$ (cf. Section 9.5), at 263 °C. These indicate a remarkable stabilization of the crystal structures in those materials stemming from the electrostatic forces between multipoles.

The electrostatic force between quadrupoles manifests itself also in the solubilities of carbon dioxide and acetylene in benzene. We consider the general cases in which only a small amount of gas resolves into a liquid, that is, the cases where Henry's law applies. Let c_0 and c be the concentrations (i.e. numbers of molecules in a unit volume) of the solute molecules in a gaseous phase and in a resolved phase; Ostwald's coefficient c/c_0 is then used as a parameter indicating the solubility. In Table 8.1 we show for various kinds of gases the ratios between their solubilities in benzene C_6H_6 and their solubilities in cyclohexane C_6H_{12} at the same temperatures. For rare gases this ratio is 0.70; for carbon dioxide and acetylene with large quadrupoles the values are much greater than that. In other words, molecules with quadrupoles are easy to resolve into benzene which has quadrupoles. The reason may be understood from the fact that the solute molecules tend to assume configurations relative to benzene molecules with reduced electrostatic energies.

8.2 Core potential combined with quadrupolar interaction

When the true shape of molecule can be well described by a parallel body of the convex core and when the strengths of the electric multipoles of the molecule are sufficiently strong, the core potential combined with those multipoles are expected to represent a good approximation to the true intermolecular potential.

The quadrupole moments Q of a symmetric diatomic molecule and a symmetric

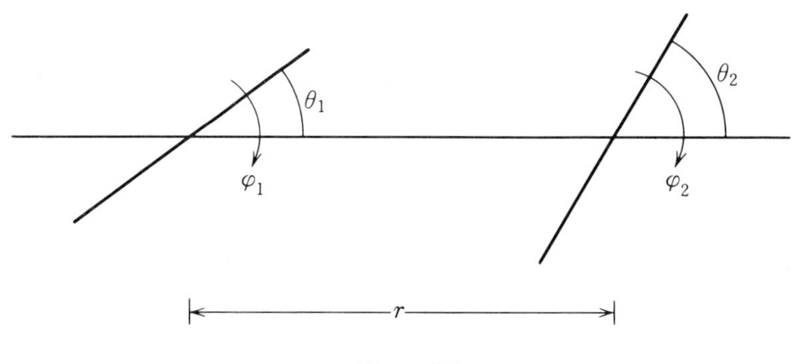

Figure 8.2

linear molecule are defined by (3.2), that is,

$$Q = \frac{1}{2}\int (2\zeta^2 - \xi^2 - \eta^2)\rho d\tau \tag{8.1}$$

Here $\rho d\tau$ denotes the electric charge in an infinitesimal volume element $d\tau$ at the position (ξ, η, ζ), the ζ axis is taken along the molecular axis, and the origin is chosen at the centre of the molecular symmetry. The values of Q have been listed in Table 3.5.

When the centres of two such molecules are separated at a distance r and the molecular axes are oriented as shown in Figure 8.2, the electrostatic energy of interaction between the qudrupoles is given by

$$\frac{3}{4}\frac{Q^2}{r^5} f(\theta_1, \theta_2, \varphi_1 - \varphi_2),$$

$$f(\theta_1, \theta_2, \varphi_1 - \varphi_2) = 1 - 5\cos^2\theta_1 - 5\cos^2\theta_2 - 15\cos^2\theta_1\cos^2\theta_2 + 2[4\cos\theta_1\cos\theta_2 - \sin\theta_1\sin\theta_2\cos(\varphi_1 - \varphi_2)]^2. \tag{8.2}$$

Here θ_1 and θ_2 are the angles of the two molecular axes with respect to the line connecting the centres; φ_1 and φ_2 are the azimuthal angles of the molecular axes around that line.

Problem

Consider two quadrupoles with momenta Q_1 and Q_2 which are formed by linear charge distributions. When these quadrupole momenta are situated as shown in Figure 8.2 at a central distance, r, we may set the interaction energy as

$$\frac{3}{4}\frac{Q_1 Q_2}{r^5} f(\theta_1, \theta_2, \varphi_1 - \varphi_2).$$

Show that the function f agrees with (8.2).

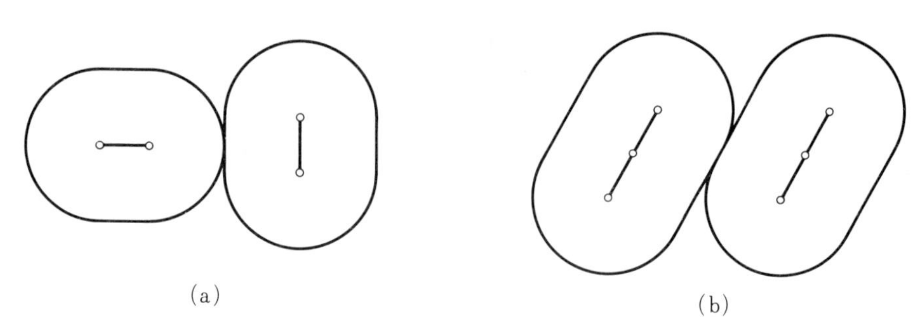

Figure 8.3 Two stable molecular orientations in the core potential combined with quadrupole interaction

Solution

We consider the charge e_1 in the quadrupole 1 located at a distance ζ_1 from the centre and the electric charge e_2 in the quadrupole 2 located at a distance ζ_2 from the centre. Denoting the distance between e_1 and e_2 by R, we have

$$R^2 = r^2 + 2r(\zeta_2 \cos\theta_2 - \zeta_1 \cos\theta_1) + \zeta_1^2 + \zeta_2^2 - 2\zeta_1\zeta_2[\cos\theta_1 \cos\theta_2 + \sin\theta_1 \sin\theta_2 \cos(\varphi_1 - \varphi_2)].$$

Expanding $e_1 e_2/R$ in a power series of r^{-1}, we may confirm that the coefficient of $e_1 e_2 \zeta_1^2 \zeta_2^2 r^{-5}$ is $(3/4)f$. Taking account of the charge distribution, we find that $\Sigma e_1 \zeta_1^2$ is equal to Q_1; and $\Sigma e_2 \zeta_2^2$ is equal to Q_2.

Adding such an interaction between the quadrupoles to the core potential, we obtain the intermolecular potential,

$$U = U_0\left[\left(\frac{\rho_0}{\rho}\right)^{12} - 2\left(\frac{\rho_0}{\rho}\right)^6\right] + \frac{3}{4}\frac{Q^2}{r^5} f(\theta_1, \theta_2, \varphi_1 - \varphi_2), \tag{8.3}$$

where ρ is the shortest distance between the cores and r is the separation between the centres.

The configuration with the lowest energy derivable from (8.3) depends on the shape of the molecules. For molecules with a shape relatively close to a sphere, such as H_2 and N_2, the configuration of Figure 8.3a has the lower energy; for relatively long and slim molecules such as CO_2, the configuration of Figure 8.3b has the lower energy.

Now it would be natural to be concerned with the accuracy of (8.3), especially the accuracy of $-2U_0(\rho_0/\rho)^6$ representing the attractive part of the van der Waals force. Let us therefore treat in the next section the problem of determining that part as accurately as possible for H_2, N_2, O_2, and CO_2.

8.3 Intermolecular potentials for H_2, N_2, O_2, and CO_2

The term $-2U_0(\rho_0/\rho)^6$ representing the van der Waals attractive force in (8.3) corresponds to the potential of dispersion force; to improve its accuracy, we must return to the rigorous expression.

We prepared for this work on the potential of dispersion force in Section 4.5. Writing such a potential as $-\mu r^{-6}$ with r representing the central distance, we found that μ is given by (4.33). In the cases of molecules of the same kind, that expression is somewhat simplified and takes the form:

$$\mu = \frac{\hbar}{3\pi}\int_0^\infty [2\alpha_\perp(i\omega) + \alpha_\parallel(i\omega)]^2 d\omega + (3\cos^2\theta_1 + 3\cos^2\theta_2 - 2)$$
$$\times \frac{\hbar}{6\pi}\int_0^\infty [2\alpha_\perp(i\omega) + \alpha_\parallel(i\omega)]\,[\alpha_\parallel(i\omega) - \alpha_\perp(i\omega)]\,d\omega$$
$$+ [(\sin\theta_1 \sin\theta_2 \cos(\varphi_1 - \varphi_2) - 2\cos\theta_1 \cos\theta_2)^2 - \cos^2\theta_1 - \cos^2\theta_2]$$
$$\times \frac{\hbar}{2\pi}\int_0^\infty [\alpha_\parallel(i\omega) - \alpha_\perp(i\omega)]^2 d\omega.$$

Here $\alpha_\parallel(i\omega)$ and $\alpha_\perp(i\omega)$ are the components of the polarizability tensor parallel and perpendicular to the molecular axis; they represent the values along the imaginary axis on the complex ω plane.

Now the ratio between the three integrals

$$\int_0^\infty \alpha_\parallel(i\omega)^2 d\omega : \int_0^\infty \alpha_\parallel(i\omega)\alpha_\perp(i\omega) d\omega : \int_0^\infty \alpha_\perp(i\omega)^2 d\omega$$

may be regarded as close to the ratio

$$\alpha_\parallel{}^2 : \alpha_\parallel\alpha_\perp : \alpha_\perp^2$$

constructed out of the components $\alpha_\parallel$ and $\alpha_\perp$ of the electrostatic polarizability. (For H_2 molecules, the former ratio takes on 1:0.746:0.560 referring to the last part of Section 4.5, while the latter takes on 1:0.769:0.591 according to Table 3.2.) Let us therefore suppose that the potential W of the van der Waals attractive force is approximated by the formula,

$$W = -W_0 F(\theta_1, \theta_2, \varphi_1 - \varphi_2) r^{-6}, \tag{8.4}$$
$$F(\theta_1, \theta_2, \varphi_1 - \varphi_2) = 2(2\alpha_\perp + \alpha_\parallel)^2 + (3\cos^2\theta_1 + 3\cos^2\theta_2 - 2)(\alpha_\parallel - \alpha_\perp)$$
$$\times (2\alpha_\perp + \alpha_\parallel) + 3[(\sin\theta_1 \sin\theta_2 \cos(\varphi_1 - \varphi_2)$$
$$- 2\cos\theta_1 \cos\theta_2)^2$$
$$- \cos^2\theta_1 - \cos^2\theta_2](\alpha_\parallel - \alpha_\perp)^2,$$

where W_0 is assumed to be a constant independent of the molecular orientation.

Although the expression (8.4) takes into account a certain amount of the dependence on the directions of the two molecules, it still fails to take into consideration an important aspect of it. Originally the factor r^{-6} in that formula derives from the square of the interaction operator proportional to r^{-3}. This operator corresponds to the energy between the dipoles of two molecules; the positions of the dipoles are here taken at the centres of the molecules. This, however, does not represent a good approximation unless the central distance r is extremely large.

Let us consider a symmetric diatomic molecule situated in a static electric field or in an electric field oscillating slowly. At the centre of the molecule a dipole, an octopole, etc. are induced by the electric field. Such an octopole is equivalent to one which would be produced by splitting the dipole into two equal parts and locating them at appropriate positions on the molecular axis. (The two positions do not significantly depend on the direction of the electric field. This may be understood from the fact that in the case of a dielectric substance with a prolate ellipsoidal shape, such positions are totally independent of the direction of the electric field.) In the cases of H_2, N_2, or O_2 molecules, those appropriate two positions are located somewhat inside the positions of the atomic nuclei. Consequently let us adopt the two ends of a rod core of such a molecule as those positions.

In the case of CO_2 molecules in an electric field, we may equally divide the induced dipole into three parts and place one each at both ends and the centre of the core; this arrangement enables us to take an accurate account of the effects up to the induced octopoles.

Thus far we have been concerned with a single molecule under the influence of a uniform electric field. The process of taking the octopolar effects into account through replacement of the dipole situated at the molecular centre by a set of small dipoles distributed along the molecular axis, however, can be likewise applied to those dipoles included in the mutual interaction between two molecules.

We first consider a diatomic molecule with a core length l. For a configuration with $\theta_1 = \theta_2 = 0$, the r^{-3} in the mutual interaction operator may be replaced by

$$\frac{1}{4}\left(\frac{2}{r^3} + \frac{1}{(r-l)^3} + \frac{1}{(r+l)^3}\right),$$

and the r^{-6} in (8.4) by its square, i.e.

$$\frac{1}{(r^2 - 2l^2)^3}\left(1 - \frac{7l^6}{r^6} + \ldots\right),$$

or to a good approximation by $(r^2 - 2l^2)^{-3}$. In general

$$W = -W_0 F(\theta_1, \theta_2, \varphi_1 - \varphi_2)[r^2 - \tfrac{1}{2}l^2(3\cos^2\theta_1 + 3\cos^2\theta_2 - 2)]^{-3}$$

substantially improves over (8.4). We adopt this form in place of $-2U_0(\rho_0/\rho)^6$ in (8.3).

In the case of CO_2 molecules with a core length l, we have

$$W = -W_0 F(\theta_1, \theta_2, \varphi_1 - \varphi_2) r^{-2}[r^2 - \tfrac{1}{2}l^2(3\cos^2\theta_1 + 3\cos^2\theta_2 - 2)]^{-2}.$$

The unknown parameter W_0 can be determined in such a way that the measured values of the second virial coefficient agree with

$$B(T) = \int_0^\infty \left\langle 1 - \exp\left(\frac{-U(\rho, \theta_1, \theta_2, \varphi_1 - \varphi_2)}{kT}\right)\right\rangle \left(\frac{\pi}{4}l^2 + 2\pi l\rho + 2\pi\rho^2\right) d\rho$$

as closely as possible. Here the intermolecular potential U is to be expressed as a function of the distance ρ between the cores and the angles θ_1, θ_2, and $\varphi_1 - \varphi_2$ representing the directions of the molecular axes; and $\langle \ldots \rangle$ implies an average with respect to those angles. We remark that quantum effects are substantial for hydrogen H_2, so that the measured values of deuterium D_2 are used instead, with small corrections arising from quantum effects.

At this stage it is not difficult to prove the following theorem. When the function $g(\theta, \varphi)$ is invariant under reflection,

$$g(\theta, \varphi) = g(\pi - \theta, \varphi + \pi)$$

and when it is expressed as a linear combination of the spherical surface harmonics of the zeroth order, the second order, and the fourth order, the identity

$$\frac{1}{4\pi}\int_0^{2\pi}\int_0^{\pi} g(\theta, \varphi) \sin\theta d\theta d\varphi = \frac{2}{5}\left[\tfrac{1}{3}g(0,0) + \tfrac{1}{3}g\left(\frac{\pi}{2}, 0\right) + \tfrac{1}{3}g\left(\frac{\pi}{2}, \frac{\pi}{2}\right)\right]$$
$$+\frac{3}{5}\left[\tfrac{1}{4}g\left(\theta_0, \frac{\pi}{4}\right) + \tfrac{1}{4}g\left(\theta_0, \frac{3\pi}{4}\right) + \tfrac{1}{4}g\left(\theta_0, \frac{5\pi}{4}\right) + \tfrac{1}{4}g\left(\theta_0, \frac{7\pi}{4}\right)\right]$$

holds, where $\theta_0 = \cos^{-1}(1/\sqrt{3})$.

With the aid of this theorem, an approximate value of the average $\langle \ldots \rangle$ with respect to the orientations of the molecular axes can be obtained by a summation with appropriate weights of the values corresponding to the following nine configurations:

$[zz]$	$\theta_1 = \theta_2 = 0$
$[zx]$	$\theta_1 = 0, \theta_2 = \pi/2$
$[xx]$	$\theta_1 = \theta_2 = \pi/2, \varphi_1 - \varphi_2 = 0$
$[xy]$	$\theta_1 = \theta_2 = \pi/2, \varphi_1 - \varphi_2 = \pi/2$
$[zd]$	$\theta_1 = 0, \theta_2 = \cos^{-1}(1/\sqrt{3})$
$[xd]$	$\theta_1 = \pi/2, \theta_2 = \cos^{-1}(1/\sqrt{3}), \varphi_1 - \varphi_2 = \pi/4$
$[dd]$	$\theta_1 = \theta_2 = \cos^{-1}(1/\sqrt{3}), \varphi_1 - \varphi_2 = 0$
$[dd']$	$\theta_1 = \theta_2 = \cos^{-1}(1/\sqrt{3}), \varphi_1 - \varphi_2 = \pi/2$
$[dd'']$	$\theta_1 = \theta_2 = \cos^{-1}(1/\sqrt{3}), \varphi_1 - \varphi_2 = \pi$

where d derives from 'diagonal' (see Figure 8.4).

We thus determine the values of

$$\frac{(2\alpha_\perp + \alpha_\parallel)^2 W_0}{(\rho_0 + l/2)^6 U_0}$$

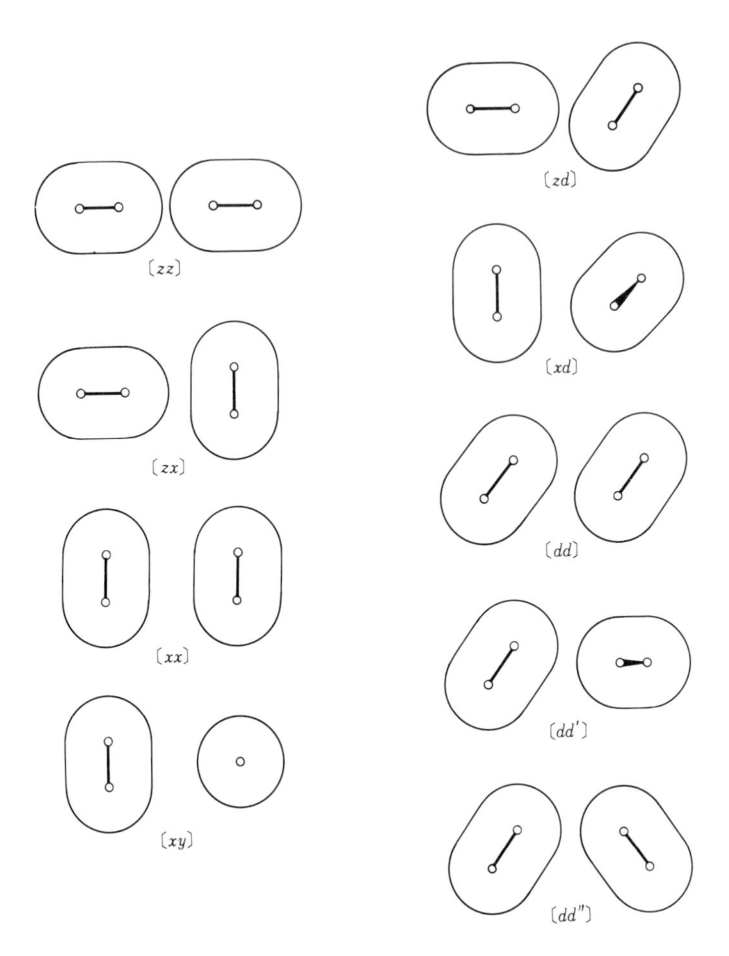

Figure 8.4 Nine typical orientations of molecules

in the following way: D_2 (and H_2) 0.90, N_2 0.88, O_2 0.84, CO_2 0.63. The electrostatic polarizabilities, $\alpha_\parallel$ and $\alpha_\perp$, are listed in Table 3.2; the core lengths l are included in Table 7.3 as $\pi l \equiv M$, together with ρ_0 and U_0.

Figures 8.5–8.8 depict the potential energies in the nine configurations in Figure 8.4 as functions of the central distance r. The configuration with the deepest potential depth is [zx] for D_2 (and H_2), and [dd] for CO_2; these agree with Figure 8.3. In other words, the core potential combined with quadrupolar interaction makes a good approximation for those molecules. The potential depth of O_2, which has almost zero quadrupole, depends little on the molecular orientations; the [xx] configuration, in which molecules are aligned side by side, represents the state of the lowest energy.

As we have seen, the ways in which the depth of the intermolecular potential

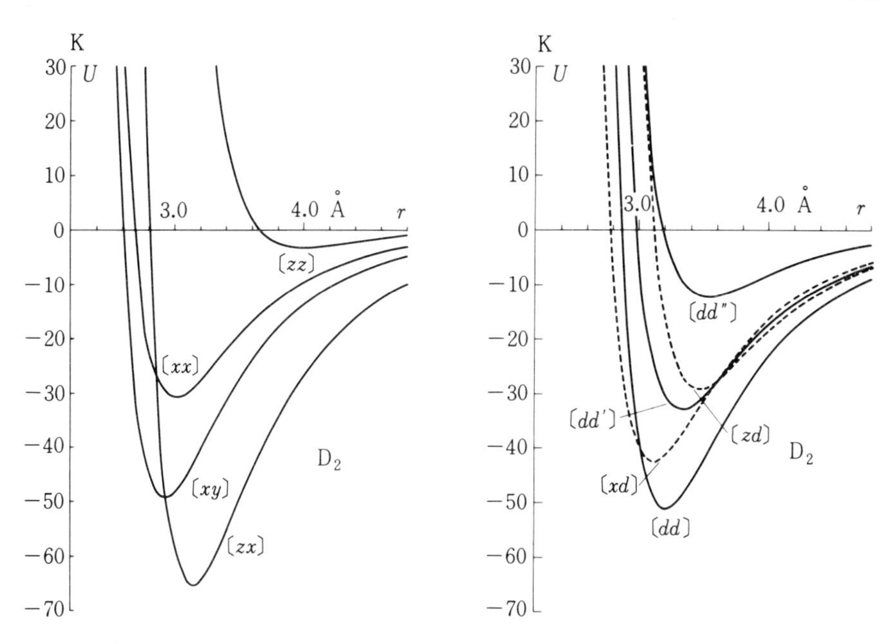

Figure 8.5 Potential between deuterium molecules at various orientations. r is the distance between the centres; U is expressed in units of degrees Kelvin after dividing by the Boltzmann constant

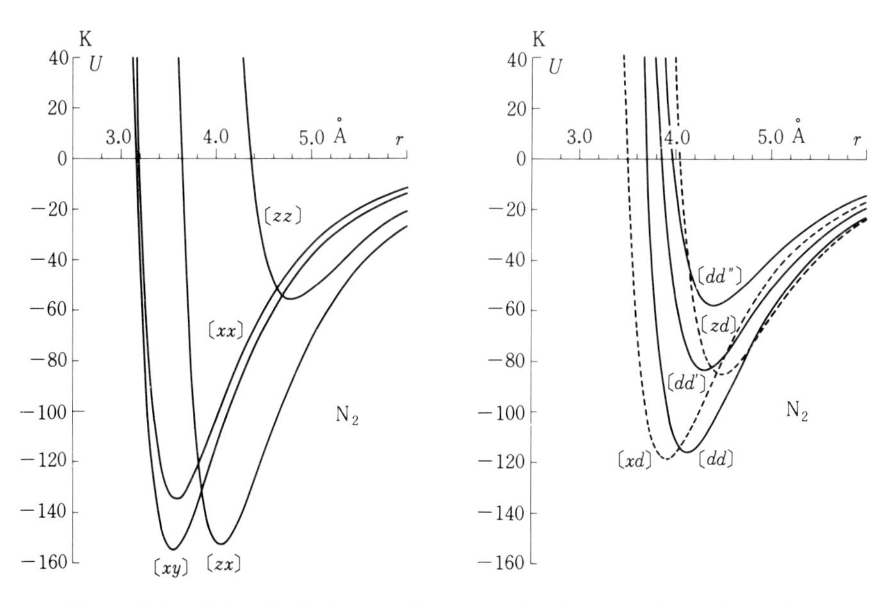

Figure 8.6 Potential between nitrogen molecules at various orientations

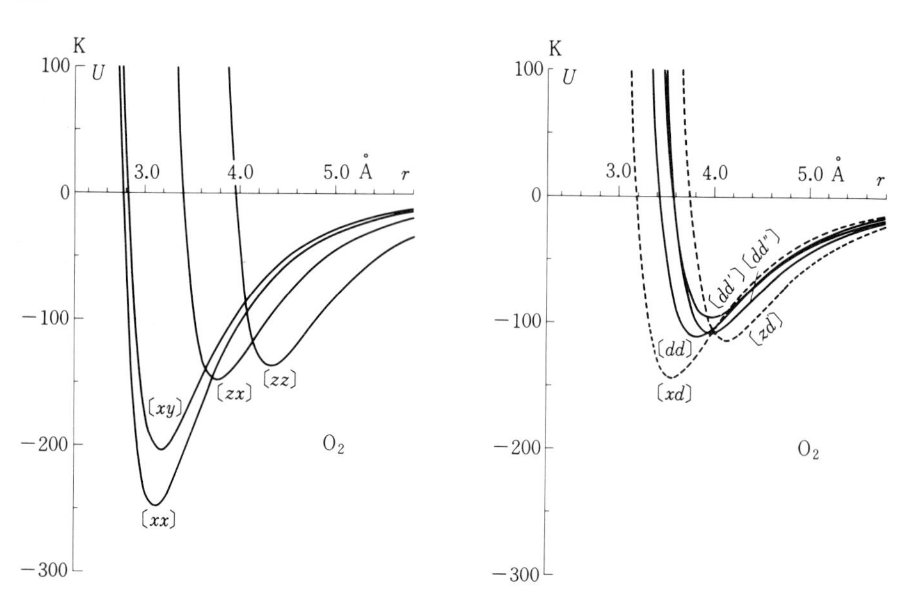

Figure 8.7 Potential between oxygen molecules at various orientations

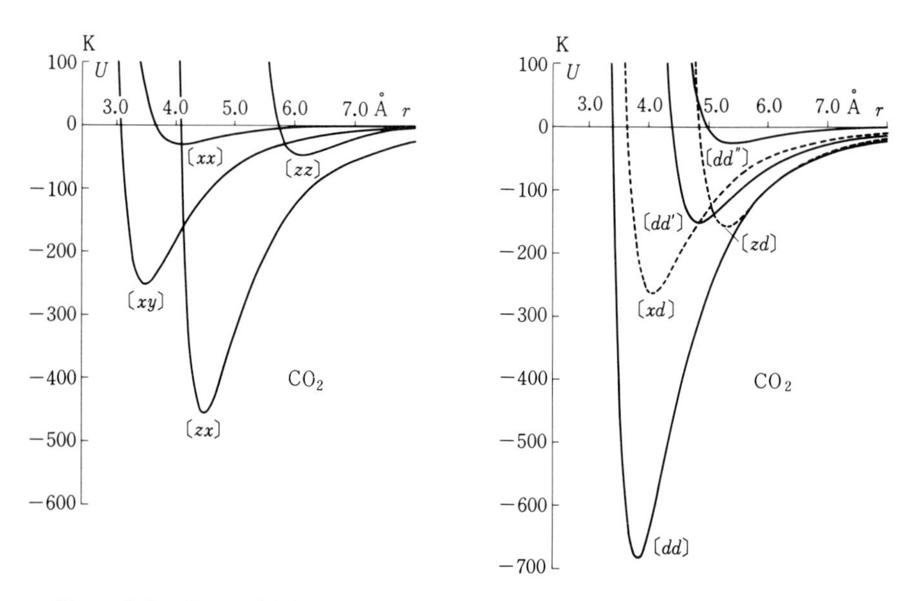

Figure 8.8 Potential between carbon dioxide molecules at various orientations

depends on the molecular orientations are quite diverse, and this fact together with the molecular shapes constitutes the origin of individuality in the intermolecular forces. The area in which such individuality manifests itself most drastically is the structures of molecular crystals.

Chapter 9

Molecular Models Representing Crystal Structures

9.1 The symmetry of crystal structures

Figure 9.1 depicts the crystal structures of Cl_2, Br_2, and I_2, namely the heavy halogens. The molecules are arranged in such a way as to make closest contact with each other, with their axes parallel to the page. The centres of the molecules here constitute an orthorhombic crystal. Since molecules in general do not necessarily have centres, however, 'a lattice constructed by the centres' has almost no meaning. Hence we consider those molecules in the same directions. To be precise, we pay attention to all those molecules which can be superposed from one to the other by parallel translations.

Such a lattice constructed by those molecules which can be superposed through parallel translations (that is, without rotation or mirror reflection) is called the *Bravais lattice.* Altogether, fourteen kinds of Bravais lattice exist (see Figure 9.2). Here P, F, I, C, and R derive from the first letters of primitive, face-centred, innenzentrum in German, C-base centred, and rhombohedral; as we shall describe shortly, these are used as a part of the symbols representing the space group. Figure 9.1 is an example of a base centred orthorhombic lattice.

Generally for a crystal structure, transformations in which a molecular configuration is superposed as a whole onto the original configuration are called the *symmetry transformations.*

When a mirror reflection with respect to a plane is involved in the symmetry transformations, this plane is called the *mirror plane*, and denoted by the symbol m.

When a rotation through an angle $2\pi/n$ around an axis is included in the symmetry transformations, that axis is called an axis of rotational symmetry of the nth order, or an *n-fold axis*; notation 2, 3, . . . , is used corresponding to $n = 2$, 3, ... When a combination of a rotation through an angle $2\pi/n$ around an axis and an inversion with respect to a point on the axis is included in the symmetry transformations, that axis is called a *rotary-inversion axis* of the nth order; corresponding to $n = 1, 3, 4$ etc., notation $\bar{1}$, $\bar{3}$, $\bar{4}$ etc. is used ($\bar{2}$ is equivalent to a mirror plane).

When a combination of a translation through a certain distance parallel to a plane and a mirror reflection with respect to the plane is included in the symmetry

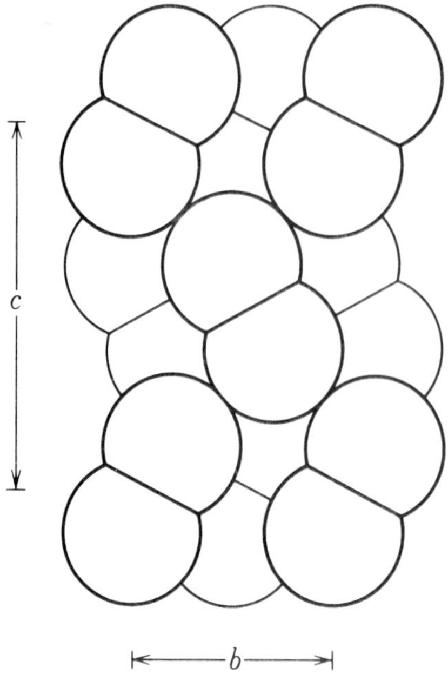

Figure 9.1

transformations, that plane is called a *glide-reflection plane.* When the parallel translation is in the direction of the a-axis and the translational distance is equal to a half of the period, it is then called the a glide-reflection plane and denoted by the symbol a; the b glide-reflection plane and the c glide-reflection plane are similarly defined. When the parallel translation is in the direction of the diagonal line of a face and its translational distance is equal to 1/2 or 1/4 of the diagonal period, it is called the diagonal glide-reflection plane or the diamond glide-reflection plane. The former appears in a netted pattern; it is thus denoted by the symbol n in connection with net; the latter appears in the crystal structure of diamond and is denoted by the symbol d.

When a combination of a rotation through an angle $2\pi/n$ around an axis and a parallel translation along the axis through a distance p/n ($p = 1, 2, \ldots, n-1$) is included in the symmetry transformations, that axis is called a *screw axis* of the nth order; the values n_p, 2_1, 4_2, etc. are used for designation.

The *space group* completely represents the symmetries of the crystal structures. The space group of the structures in Figure 9.1 is denoted by *Cmca.* This symbol implies that the Bravais lattice is a C-base centred, orthorhombic lattice and the crystal has a mirror plane perpendicular to the a axis, a c glide-reflection plane perpendicular to the b axis, and an a glide-reflection plane perpendicular to the c axis. In general the first capital letter of the designation distinguishes whether the Bravais lattice is primitive (P), body centred (I), etc.; the symmetry elements then

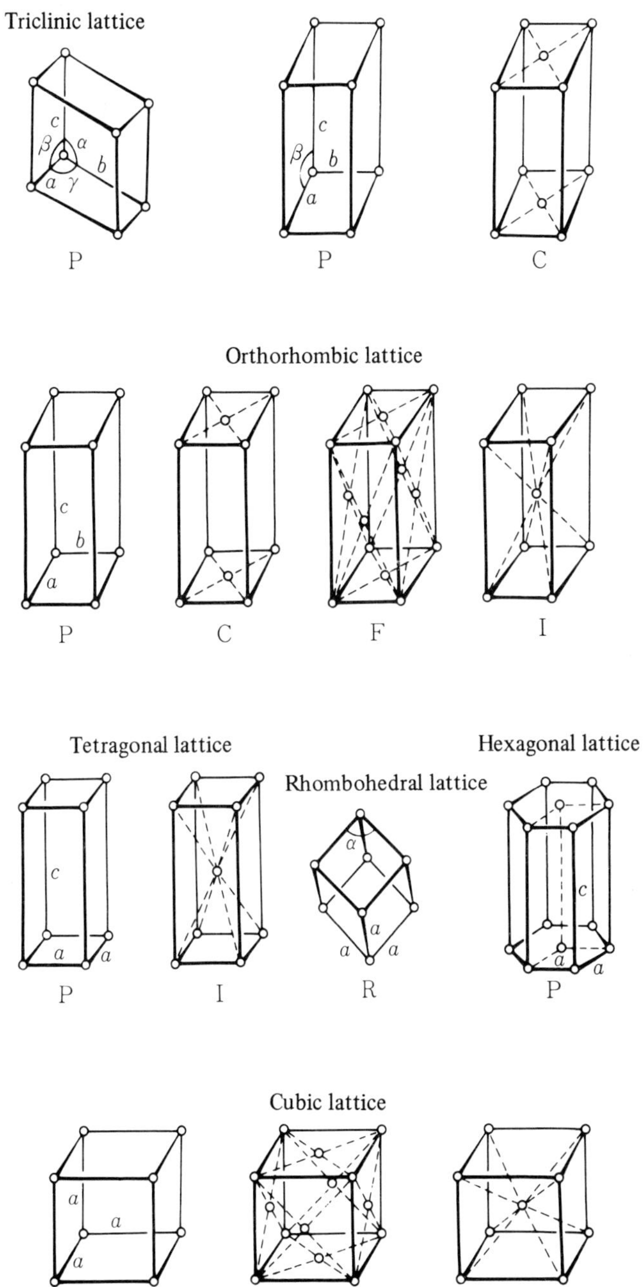

Figure 9.2 Bravais lattices

Table 9.1 Principal space groups. In parentheses are Schoenflies' symbols

monoclinic crystal $P2_1/c(C_{2h}^5)$, $C2/c(C_{2h}^6)$

$P2_1/c$: a screw axis of the second order in the direction of the *b* axis, and a *c* glide-reflection plane perpendicular to it.

orthorhombic crystal $Pnnm(D_{2h}^{12})$, $Pbca(D_{2h}^{15})$, $Pnma(D_{2h}^{16})$, $Cmca(D_{2h}^{18})$

$Pnma$: a diagonal glide-reflection plane, a mirror plane, and an *a* glide reflection plane perpendicular to the *a*, *b*, and *c* axes, respectively.

tetragonal crystal $P4_2/nmc(D_{4h}^{15})$

$P4_2/nmc$: a screw axis of the fourth order in the direction of the *c* axis, a diagonal glide-reflection plane perpendicular to it, a mirror plane perpendicular to the *a* axis, and a *c* glide-reflection plane in the diagonal plane containing the *c* axis.

trigonal crystal $R\bar{3}(C_{3i}^2)$, $P\bar{3}m1(D_{3d}^3)$

$P\bar{3}m1$: Bravais lattice is simple-hexagonal; a rotary-inversion axis of the third order in the direction of the principal axis, a mirror plane perpendicular to the *a* axis; no symmetry elements in the diagonal plane containing the principal axis.

cubic crystal $Pa3(T_h^6)$, $I\bar{4}3m(T_d^3)$

$I\bar{4}3m$: a rotary-inversion axis of the fourth order along the axial direction, a three-fold axis along the diagonal direction, and a mirror plane parallel to the diagonal plane containing an axis.

follow according to the conventionally agreed sequence. In Table 9.1 we summarize the space groups used in this chapter, together with concise explanations.

9.2 Representation of *Pa*3 structures by means of quadrupole spheres

Those molecules with strong electric quadrupoles whose shapes do not differ significantly from the sphere frequently assume a structure of the cubic system belonging to the space group *Pa*3. A question then arises as to how we may understand and explain this fact and a series of facts closely related to it; the purpose of this chapter is to provide an answer to this question through representation by means of molecular models.

The force between two electric quadrupoles may become attractive or repulsive depending on their relative configuration. That is, the force between quadrupoles is a sensitive function of the molecular configuration. This is an aspect which differs significantly from the ordinary van der Waals attractive force which simply acts to attract the molecules together. It is then expected that the configuration of nearly spherical quadrupolar molecules in a crystal is determined by the quadrupole force. To substantiate such an expectation, we begin with production of many spheres with quadrupoles.

In 1960 ferrite magnets were first widely produced; magnets of arbitrary sizes became readily available, especially in a disc or in a cylindrical shape. A sphere with

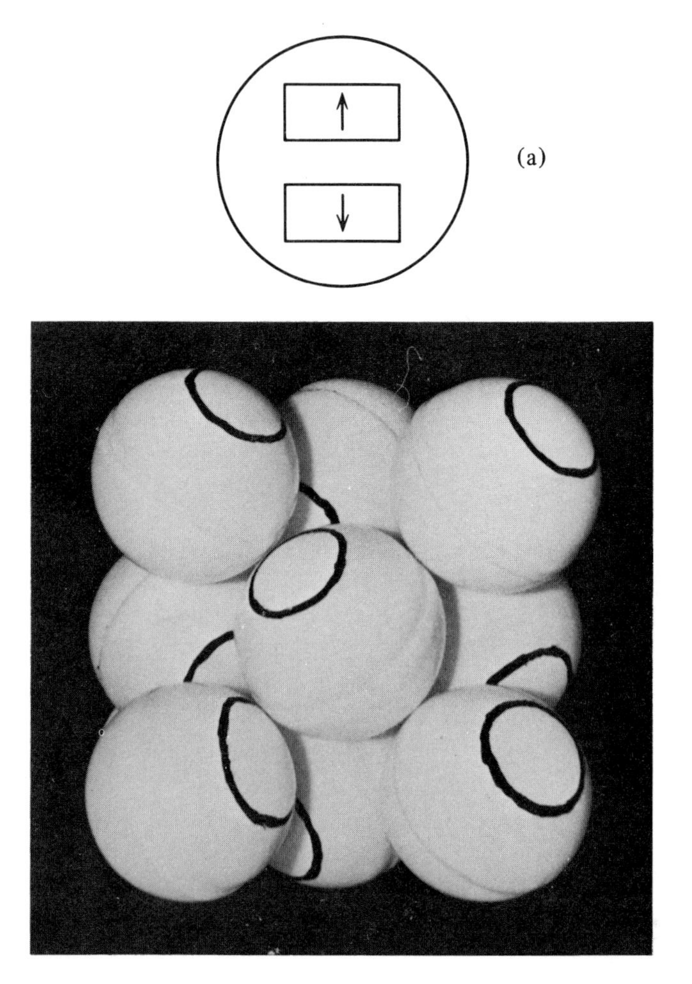

Figure 9.3 (a) Sphere with a quadrupole produced by fixing two ferrite magnets inside a ping-pong sphere; (b) a crystal structure constructed from such spheres. From T. Kihara (1960), *J. Phys. Soc., Japan,* **15** 1920

a quadrupole is produced by fixing two magnets inside a ping-pong sphere as shown in Figure 9.3a. We choose the weight of the magnet in such a way that the specific weight of the sphere is exactly equal to unity. When several such spheres are placed in a water pool, gravity is cancelled and a crystal structure appears; Figure 9.3b is such a structure (T. Kihara (1960), *J. Phys. Soc., Japan,* **15**, 1920).

It is clear that this structure belongs to the space group P*a*3 actually observed. The quadrupole axis, that is, the orientation of the molecular axis, assumes one of the four directions; molecules oriented in the same direction form a simple cubic lattice. An *a* glide-reflection plane exists perpendicular to an axis of the cubic lattice; a three-fold axis, in a diagonal direction.

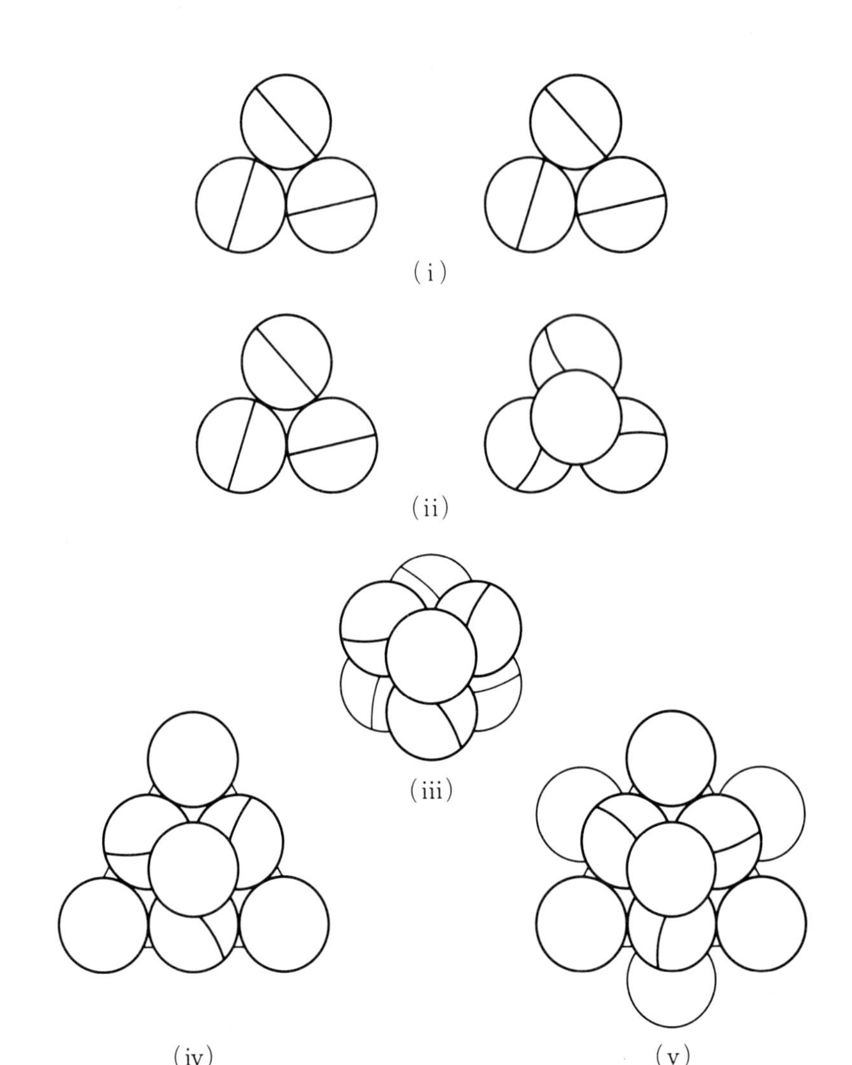

Figure 9.4 Example of the order of assembling 14 quadrupole spheres. (i) Make two assemblies of three spheres on a desk. (ii) Let a sphere with its axis in the vertical direction approach the centre of one of the two assemblies from above; the three spheres underneath then change their orientations slightly to form a stable assembly of four spheres. (iii) When this assembly is placed above the other assembly, the three spheres underneath also change their orientations slightly. (iv) Add three spheres with their axes in the vertical direction around the three spheres above. (v) Turn the entire assembly upside down and add four spheres similarly. The final figure shows the P*a*3 structure observed from the direction of the three-fold axis

Currently, magnets with shapes other than a disc can be easily produced; a quadrupole sphere may be produced by binding two hemispherical magnets. In these cases, since the magnetic forces are substantially strong, the gravity is negligible. Structures similar to the above can be reproduced by assembling those spheres on the desk.

These structures, once produced, are sufficiently stable. It is desirable, however, to follow an appropriate order when assembling the spheres. Figure 9.4 shows such an example.

9.3 Crystals of prolate, uniaxial quadrupolar molecules

In the previous chapter we investigated how the potential depth between D_2 molecules or between CO_2 molecules depends on the relative orientations of those molecules. According to the result, the configuration with the lowest potential energy between two D_2 molecules is that as shown in Figure 8.3a. In contrast, the most stable configuration between two CO_2 molecules assumes a slanted parallel orientation as shown in Figure 8.3b.

Model 1 and Model 2 in Figure 9.5 have those properties appropriate to the D_2 molecules and the CO_2 molecules, respectively. Model 1 is a slightly elongated version of the quadrupole spheres in the previous section; both ends of model 2 are made of plastic. The stable shapes of assemblies of three such model spheres have the same symmetries as the configurations of the three quadrupole spheres treated in the previous section. Assemblies of many such models thus produce $Pa3$ structures (Figures 9.6 and 9.7). In fact, for crystals of deuterium D_2, the structure at sufficiently low temperatures (below 2.5 K) is known to be such a $Pa3$. The crystal structure of carbon dioxide has been known some time; the foregoing structure produced by the molecular models agrees with it.

As a sphere with a quadrupole is gradually elongated in an ellipoidal shape, there is a point at which the $Pa3$ structure becomes unstable. This boundary corresponds to Model 3 of Figure 9.5. In this model, two kinds of lattice structure may be produced with a similar stability. One is $Pa3$ of the cubic system; the other belongs to the orthorhombic system and assumes a $Pnnm$ structure shown in Figure 9.8. The former looks like Figure 9.9 from the direction of a diagonal line of a face or from the (110) direction; it clearly shows a correspondence with the $Pnnm$ structure. Through application of forcible change in directions of a few molecular models, the structures of Figures 9.8 and 9.9 make 'transitions' from one to the other.

The crystals of acetylene HC≡CH make a phase transition at −140 °C and assume a $Pa3$ structure similar to CO_2 crystals at high temperatures. A possibility remains that the crystal structure at low temperatures may correspond to the $Pnnm$ structure described above. (According to H.K. Koski and E. Sándor (1975), *Acta Cryst.*, B, **31**, 350, the crystal structure of C_2D_2 at 4.2 K has been determined to be $Cmca$, $Z = 4$, by means of neutron diffraction. The author is grateful to Dr Koski

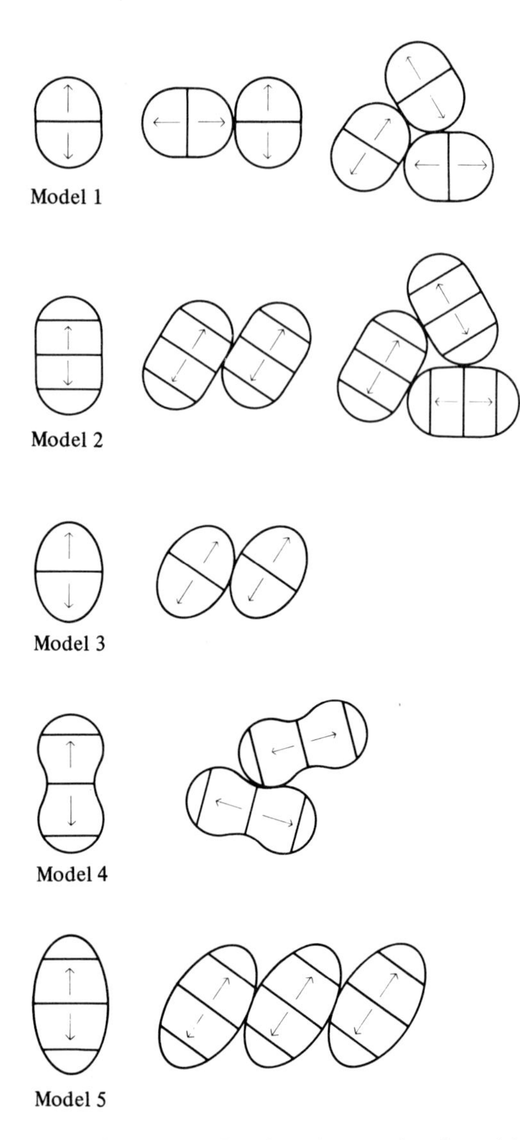

Figure 9.5 Models of prolate molecules with uniaxial quadrupoles. The parts without arrows are made of plastic

for calling his attention to this. Incidentally, the crystal structure of ethylene C_2H_4 is P*nnm* with respect to the positions of C atoms.)

The molecule of cyanogen N≡C–C≡N may have a shape slightly concave around its middle part. Model 4 of Figure 9.5 describes the shape and polarity of this molecule together with a stable configuration of two such molecules assembled close to each other. Figure 9.10 shows a stable configuration when models of such

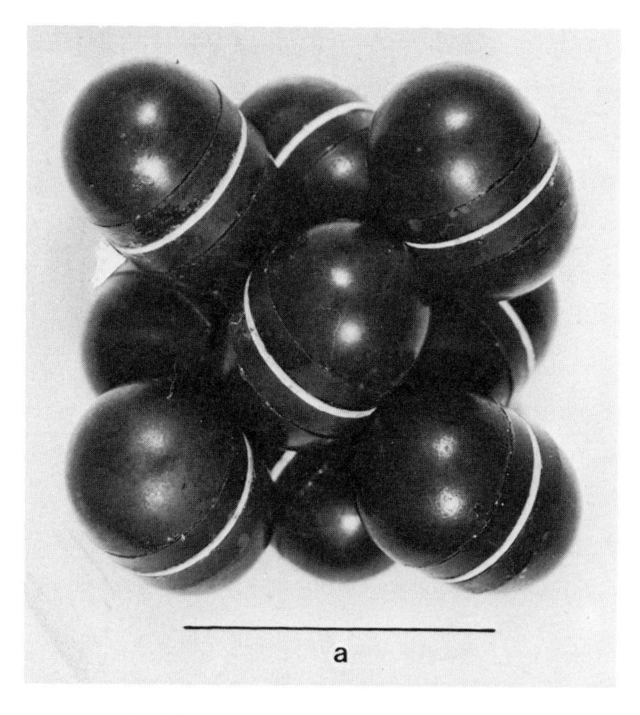

Figure 9.6 The crystal structure of deuterium at low temperatures simulated by Model 1

Figure 9.7 The crystal structure of carbon dioxide simulated by Model 2

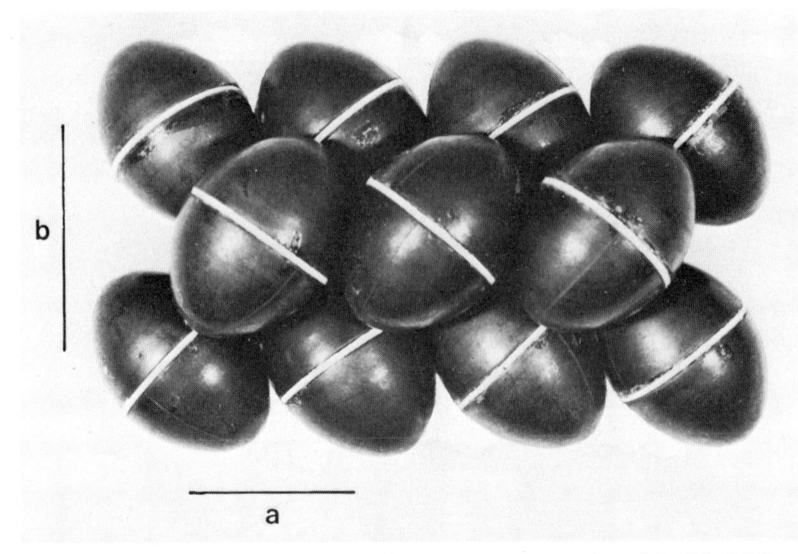

Figure 9.8 The orthorhombic P*nnm* structure produced by Model 3

molecules are assembled two-dimensionally on the desk; Figure 9.11, constructed by piling up such layers in alternate directions, corresponds to a stable three-dimensional structure. This is a P*bca* structure belonging to the orthorhombic system, and agrees with the actual crystal structure of cyanogen.

A symmetric two-atom substitution product of ethane, 1,2-dichloroethane

Figure 9.9 The cubic P*a*3 structure produced by Model 3, viewed from the (110) direction

Figure 9.10 The two-dimensional structure produced by Model 4

$ClCH_2CH_2Cl$ and a symmetric two-atom substitution product of cyclohexane, trans-1,4-dibromocyclohexane

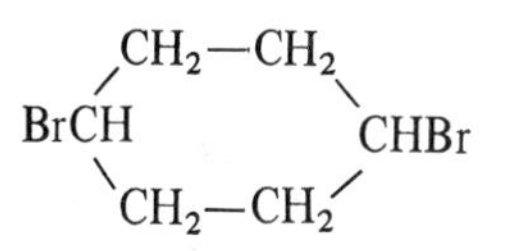

have molecular shapes which are somewhat convex around their middle parts (cf.

Figure 9.11 The crystal structure of cyanogen simulated by Model 4

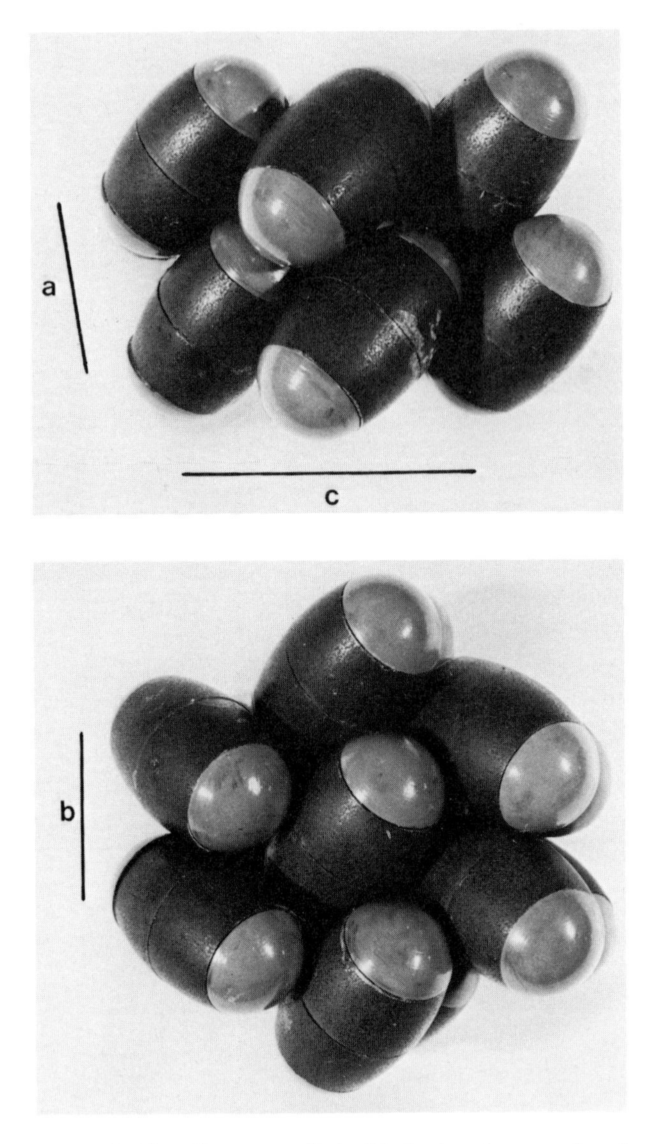

Figure 9.12 The monoclinic $P2_1/c$ structure produced by Model 5

Figure 9.13a). Model 5 shows an idealized representation of the shape and polarity of such molecules; this model produces a $P2_1/c$ structure (Figure 9.12) of the monoclinic system, which corresponds to the actual crystal structure.

Table 9.2 compares the lattice constants of actual crystals and corresponding values of the structures constructed by the models, where Z designates the number of molecules in a unit cell.

Table 9.2 Crystal structures of prolate quadrupolar molecules

$(CN)_2$	*Pbca*	$Z = 4$	$a/c = 0.874$	$b/c = 0.891$	
Model 4	*Pbca*	$Z = 4$	$a/c = 0.90$	$b/c = 0.90$	
1,2-dichloroethane					
	$P2_1/c$	$Z = 2$	$a/b = 0.860$	$c/b = 1.454$	$\beta = 103°30'(-140\ °C)$
	$P2_1/c$	$Z = 2$	$a/b = 0.906$	$c/b = 1.439$	$\beta = 109°30'(-50\ °C)$*
trans-1,4-dibromocyclohexane					
			$a/b = 1.083$	$c/b = 2.144$	$\beta = 101°\ 49'$
Model 5	$P2_1/c$	$Z = 2$	$a/b = 1.0$	$c/b = 1.75$	$\beta = 100°$

*Below −96 °C, the molecules rotate around the Cl—Cl axis.

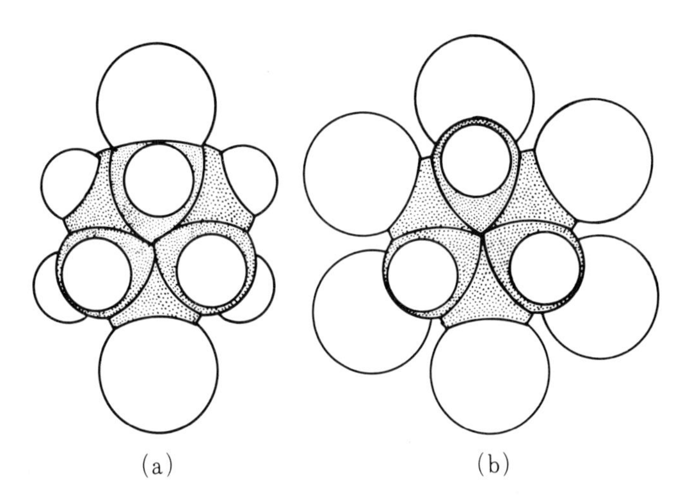

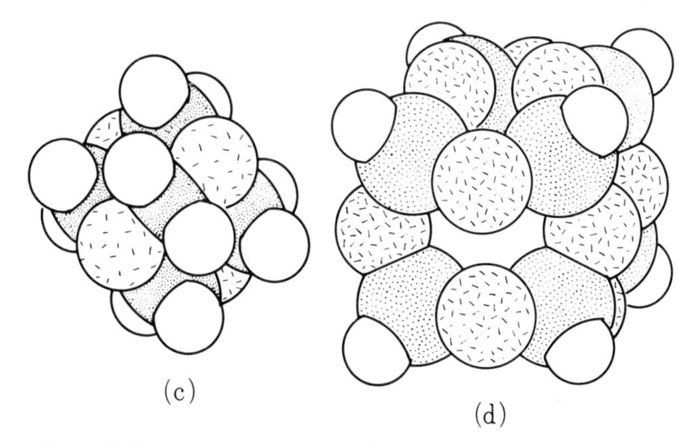

Figure 9.13 (a) trans-1,4-dichlorocyclohexane $C_6H_{10}Cl_2$. C atoms are shaded; Cl atoms, large spheres. (b) β-hexachlorocyclohexane $C_6H_6Cl_6$. (c) hexamethylenetetramine $(CH_2)_6N_4$; N atoms are shaded by dashes. (d) $(HSi)_8O_{12}$; Si atoms are shaded by dots, O atoms are shaded by dashes

In Figure 9.13 we describe the shapes of some of those multipolar molecules treated in Sections 9.3–9.6.

9.4 Crystals of oblate, uniaxial quadrupolar molecules

The molecules of cyclohexane C_6H_{12} have an almost axisymmetric shape. β-hexachlorocyclohexane $C_6H_6Cl_6$ is produced by replacing all of the six H atoms located away from the symmetry axis of cyclohexane by chlorine atoms; β-hexabromocyclohexane $C_6H_6Br_6$, by bromine atoms. The shapes of those

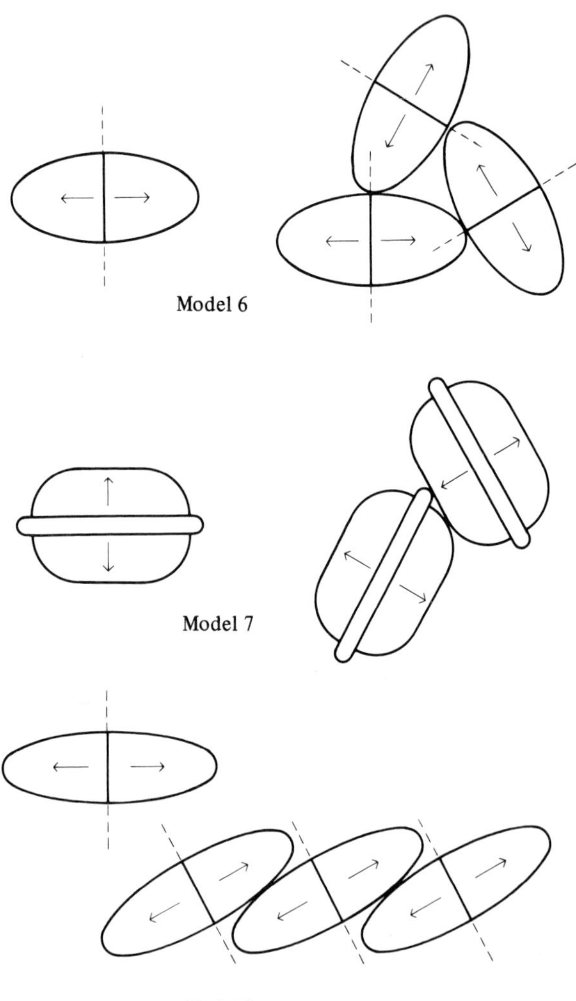

Figure 9.14 Models of oblate molecules with uniaxial quadrupoles

molecules may be approximated by substantially oblate ellipsoids (cf. Figure 9.13b). In these molecules, Cl atoms or Br atoms attract electrons so that on the molecular surfaces negative charges are accumulated far away from the molecular axis; positive charges, near the molecular axis. Electric quadrupoles are thus formed.

Model 6 of Figure 9.14 is an idealized representation of the shape and polarity of such molecules. Pieces of ferrite with the shape of one sixth of an ellipsoid are magnetized in the outward direction; six pieces are then united into an ellipsoid. An assembly of such models reproduces the actual crystal structure, that is, a *Pa*3 structure belonging to the cubic system (Figure 9.15).

Model 7 represents an axisymmetric idealization of the shape and polarity of benzene C_6H_6. The electric quadrupole owing to the π electrons has been replaced by a magnetic quadrupole made of ferrite magnets; the effect of 'protrusion' arising from the six H atoms has been taken into account by means of a thin plastic disc inserted in the middle part. Without such protrusion, this model would produce the same result as in the case of the quadrupolar spheres, that is, a *Pa*3 structure of the cubic system similar to Model 6; this would not agree with reality. The actual crystal structure is *Pbca* of the orthorhombic system with lower symmetry than that. As Figure 9.16 shows, this structure is realized by our models. Incidentally,

Figure 9.15 The cubic *Pa*3 structure produced by Model 6

Figure 9.16 The crystal structure of benzene simulated by Model 7

the fact that benzene and cyanogen have similar crystal structures is related to the similarity in the stable configurations that the two models produce.

Model 8 is a further oblate version of Model 6. (The axial ratios are 2:1 for Model 6, and 3:1 for Model 8.) An assembly of such models yields a $P2_1/c$ structure of the monoclinic system shown in Figure 9.17. The crystal structures of

naphthalene

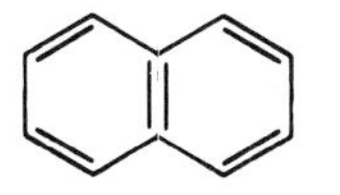

and of tetracyanoethylene

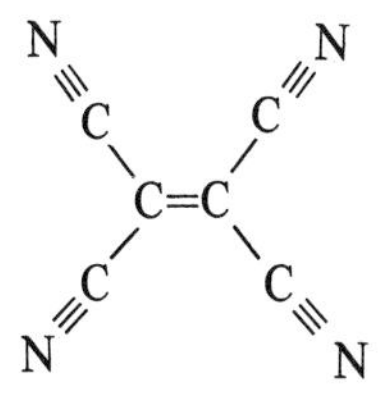

Table 9.3 Crystal structures of oblate quadrupolar molecules

C_6H_6	*Pbca*	$Z = 4$	$b/a = 0.728$	$c/a = 0.772$	(−3 °C)
Model 7	*Pbca*	$Z = 4$	$b/a = 0.8$	$c/a = 0.9$	
naphthalene	$P2_1/c$	$Z = 2$	$a/b = 1.35$	$c/b = 1.37$	$\beta = 116°$
$(CN)_2C{=}C(CN)_2$	$P2_1/c$	$Z = 2$	$a/b = 1.13$	$c/b = 1.75$	$\beta = 137°$
Model 8	$P2_1/c$	$Z = 2$	$a/b = 1.0$	$c/b = 1.4$	$\beta = 120°$

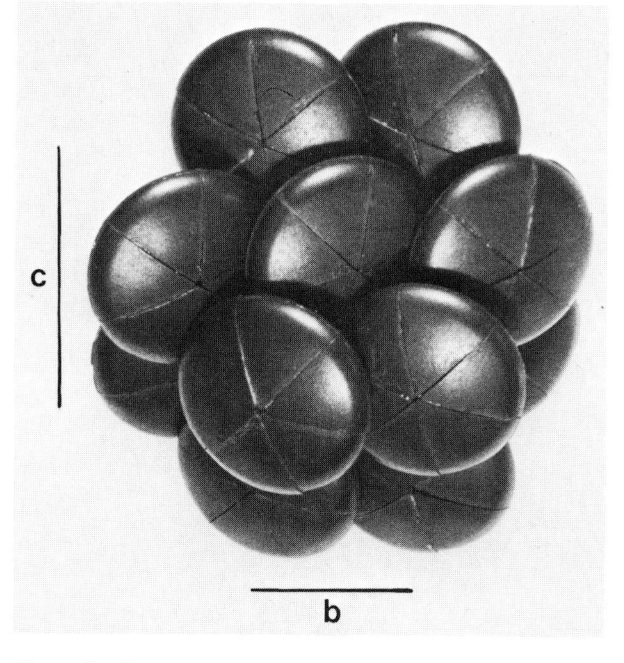

Figure 9.17 The monoclinic $P2_1/c$ structure produced by Model 8

which replaces the four H atoms in ethylene by cyano radicals, correspond to that. Table 9.3 summarizes comparison of the lattice constants.

Forces between quadrupoles play a significant part in the molecular configurations also in the 1:1 mixed crystals of benzene and chlorine Cl_2, and in similar mixed crystals of benzene and bromine Br_2. Since the sign of the quadrupole moment Q is negative for benzene and positive for Cl_2 (Table 3.5), the lowest electrostatic energy is achieved by the configuration in which the axis of the Cl_2 molecule overlaps with the six-fold axis of the benzene molecule; such a configuration is realized in the actual crystals. In the 1:1 mixed crystals of naphthalene and tetracyanoethylene the two kinds pile up alternately with their molecular planes kept parallel to each other. Such a configuration may also be understood as the configuration with the lowest electrostatic energy.

In these intermolecular interactions, a partial transfer of electrons between molecules is involved, that is, a *charge transfer complex* is formed; these interactions are of interest from such a point of view as well.

9.5 Octopolar molecules

The molecular shape of hexamethylenetetramine $(CH_2)_6N_4$ is quite close to a sphere; this molecule is constructed by placing N atoms at the vertices of a regular tetrahedron and methylene radicals at the middle points of the edges and by

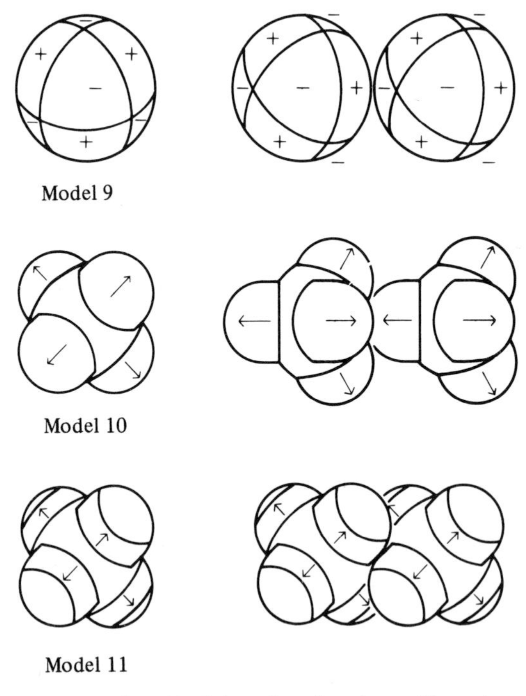

Figure 9.18 Models of molecules with octopolar symmetry

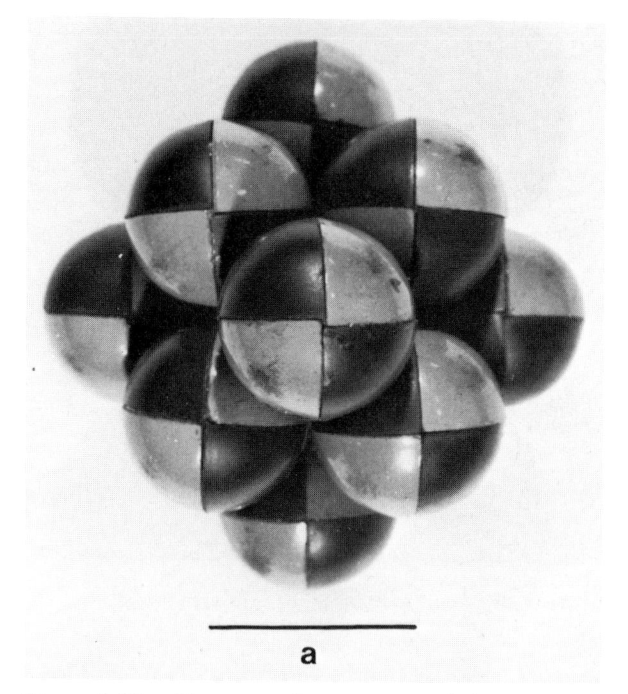

Figure 9.19 The crystal structure of hexamethylenetetramine simulated by Model 9. In this model only, either the positive or the negative poles is coloured grey

rounding off the entire shape into a sphere (see Figure 9.13c). Since the electronegativity of the N atom is substantially large, negative charges are gathered around N atoms on the surface of the sphere; this molecule thus has an electric octopole.

Such an electric octopole may be simulated by a magnetic octopole in the following way, Many pieces of ferrite with the shape of one eighth of a sphere are produced; half of them are magnetized in the outward direction, the remaining half in the inward direction. Eight such magnets are assembled with alternate directions of magnetization and are bonded together into a spherical shape. When two Models 9 so constructed approach each other, they assume the configuration shown in Figure 9.18, in which the positive and negative poles face each other. When many (15, say) of them are assembled into a most stable configuration, the molecules make mutual contact with positive and negative poles facing each other again, as shown in Figure 9.19. Since all of the molecular orientations are equivalent in such a structure, the crystal lattice constitutes the only Bravais lattice – the body centred cubic lattice. Its space group is $I\bar{4}\,3m$.

Many small crystals of hexamethylenetetramine are produced by resolving it in alcohol and then by evaporating the alcohol. Figure 9.20 shows a magnified picture of such a crystal. Figure 9.21 is a corresponding picture of simulation in which the

Figure 9.20 The crystal of hexamethylenetetramine (approximately 1.5 mm in diameter)

Figure 9.21 The crystal shape of hexamethylenetetramine simulated by an assembly of many molecular models, viewed from the same direction as in the previous figure

Figure 9.22 The crystal structure of SiF_4 simulated by Model 10

Figure 9.23 The two-dimensional structure produced by Model 11

Figure 9.24 The tetragonal crystal structure of B_4Cl_4 simulated by Model 11

Figure 9.25 The crystal structure of CF_4 at low temperatures simulated by Model 11

shape of a crystal is reproduced by assembling a sufficiently large number of molecular models.

The crystal of silicon tetrafluoride SiF_4 assumes a structure similar to a $(CH_2)_6N_4$ crystal. Since that molecule has an exeedingly large octopole arising from a substantial difference in electronegativities between an F atom and an Si atom, it may be idealized as in Model 10. Figure 9.22 shows a crystal structure simulated by assembling such models.

It is to be noted here that these crystal structures do not assume the closest packed configurations of the molecules. This may also be seen from the fact that only eight molecules contact with a given molecule. Such a feature reflects a substantial strength of molecular multipoles.

The octopole of carbon tetrafluoride CF_4 is not so strong. To see the effect, we have tentatively constructed Model 11; the configuration of two such models is then different from the case of SiF_4, as Figure 9.18 indicates. Figure 9.23 shows a stable two-dimensional configuration of such models. When such two-dimensional

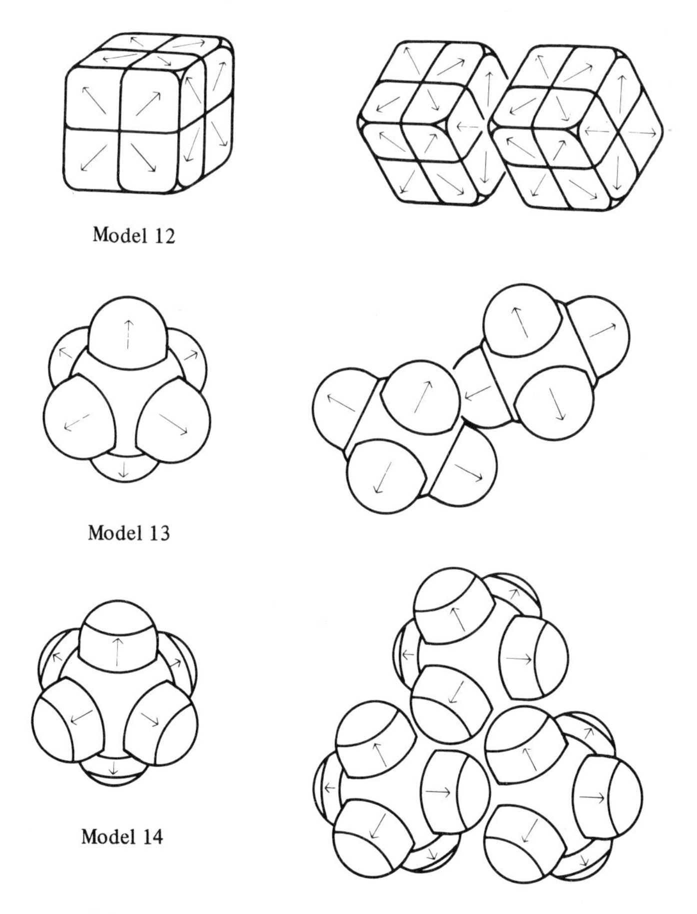

Figure 9.26 Models of molecules with cubic or octahedral symmetry

configurations are piled up into layers, two kinds of structures may be produced depending on the way of piling up. One is $P4_2/nmc$ belonging to the tetragonal system and corresponds to the crystal structure of B_4Cl_4; the other is $C2/c$ belonging to the monoclinic system and reproduces the crystal structure of CF_4 at low temperatures (D.N. Bol'shutkin, V.M. Gasan, A.I. Prokhvatilov, and A.I. Erenburg (1972), *Acta Cryst.*, B, **28**, 3542). See Figures 9.24 and 9.25, respectively.

Figure 9.27 The trigonal $R\bar{3}$ structure represented by Model 12

9.6 Molecules with octahedral symmetry

$(HSi)_8O_{12}$, which is constructed by placing SiH at the vertices of a cube and oxygen atoms at the middle points of the edges (see Figure 9.13d), is called octa(silsesquioxane); sesqui means one and a half times. This molecule and octa(methylsilsesquioxane) $(CH_3Si)_8O_{12}$ in which H in the former molecule is replaced by the methyl radical have shapes close to a cubic body; on their surfaces, excessive positive charges are found at each vertex.

Model 12 of Figure 9.26 represents an idealization of the shape and polarity of such molecules. Figure 9.27 shows an assembly of eight such molecular models; it correctly describes the actual crystal structure – an $R\bar{3}$ structure belonging to the trigonal system. In this case again, the structure does not represent a most closely packed configuration of molecules.

Uranium hexafluoride UF_6 uranium hexachloride UCl_6, and tungsten hexachloride WCl_6 have similar shapes, but the strengths of multipoles (hexadecapoles in this case) are substantially different. Clearly the multipole of UF_6 is the strongest; since the ionization energies of U and W are 4 eV and 8 eV, the multipole of WCl_6 is the weakest.

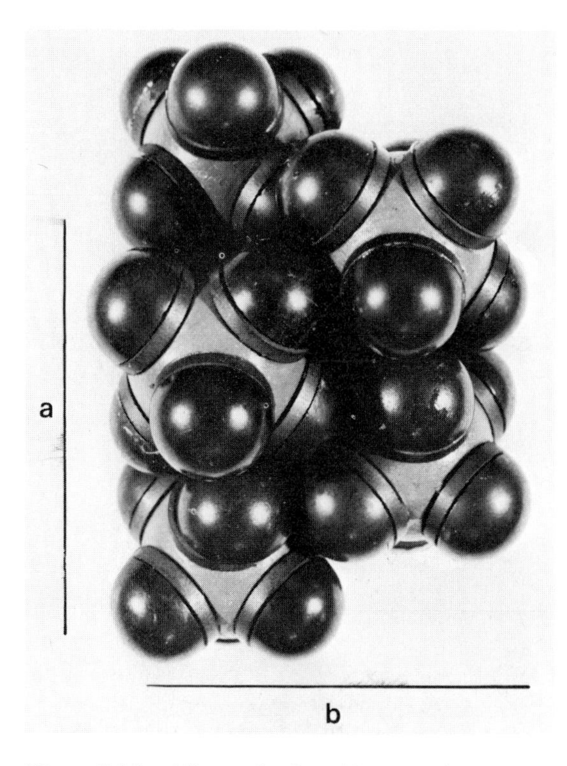

Figure 9.28 The orthorhombic crystal structure of UF_6 simulated by Model 13

Figure 9.29 The trigonal crystal structure of UF_6 simulated by Model 13

Model 13 has been constructed under the expectation of describing molecules with strong multipoles; Model 14, molecules with weak multipoles. Stable configurations of two or three models are also shown.

Two kinds of structures are constructed from Model 13. One is the crystal structure of UF_6, representing a P*nma* structure belonging to the orthorhombic system (Figure 9.28); the other is the crystal structure of UCl_6, representing $P\bar{3}m1$ belonging to the trigonal system (Figure 9.29).

These structures cannot be constructed by Model 14. It rather produces an $R\bar{3}$

Figure 9.30 The two-dimensional structure produced by Model 14

Figure 9.31 The crystal structure of WCl_6 simulated by Model 14

Table 9.4 Crystal structures of multipolar molecules

Compound	Space group	Z	Lattice constants	
$(HSi)_8O_{12}$	$R\bar{3}$	$Z = 1$	$\alpha = 76°50'$	
$(CH_3Si)_8O_{12}$	$R\bar{3}$	$Z = 1$	$\alpha = 95°39'$	
Model 12	$R\bar{3}$	$Z = 1$	$\alpha = 100°$	
UF_6	$Pnma$	$Z = 4$	$b/a = 0.905$,	$c/a = 0.526$
Model 13	$Pnma$	$Z = 4$	$b/a = 0.92$,	$c/a = 0.53$
UCl_6	$P\bar{3}m1$	$Z = 3$	$c/a = 0.551$	
Model 13	$P\bar{3}m1$	$Z = 3$	$c/a = 0.55$	
WCl_6	$R\bar{3}$	$Z = 1$	$\alpha = 55°$	
Model 14	$R\bar{3}$	$Z = 1$	$\alpha = 53°$	

structure of the rhombohedral lattice constructed by piling up the stable two-dimensional configurations of Figure 9.30 in layers; it agrees with the crystal structure of WCl_6 (Figure 9.31).

Table 9.4 summarizes comparison of lattice constants.

Chapter 10

Viscosity and Thermal Conductivity of Gases

10.1 Velocity distribution function for molecules in gases

In Chapters 5–7, we considered spatially uniform gases. When the temperature is not uniform in a gas, the phenomenon of thermal conduction takes place. When the macroscopic flow is not uniform, viscosity appears. When the density of a particular kind of molecule is not uniform, those molecules diffuse. These, respectively, are interpreted as phenomena in which thermal energy, momentum, and the molecules themselves are transported; they are thus called *transport phenomena*.

Transport of thermal energy is proportional to the temperature gradient; its constant of proportionality is the thermal conductivity. Transport of momentum is proportional to the gradient of flow velocity; its constant of proportionality is the viscosity. Transport of a given kind of molecules is proportional to the density gradient of that kind of molecules; its constant of proportionality is the diffusion coefficient. Thermal conductivity, viscosity, the coefficient of diffusion, etc. are generally called the *transport coefficients.*

Transport coefficients in gases are analysed in terms of collisions between gas molecules. Unless the density of the gas is excessively high, the probability of three or more molecules making complex collisions is small; a transport coefficient is therefore expressed as a summation of collisional effects between two molecules. Such effects of collisions are intimately related to intermolecular forces. Thus one can treat transport coefficients of gases in relation to the intermolecular forces.

In this chapter we consider a single-component gas. When a gas is not uniform in space, it is not generally stationary in time. A fundamental quantity for treatment of such a system is the velocity distribution function $f(\mathbf{c}, \mathbf{r}, t)$ of gas molecules, defined in the following way. The number of molecules expected in an infinitesimal volume $d\mathbf{r} \equiv dxdydz$ at the position $\mathbf{r}$ and in an infinitesimal domain $d\mathbf{c} \equiv dc_x dc_y dc_z$ around the velocity $\mathbf{c}$ at time t is

$$f(\mathbf{c}, \mathbf{r}, t)d\mathbf{c}d\mathbf{r}.$$

The number of molecules in a unit volume, namely, the *number density* is given by the integral of f over the entire domain of the molecular velocities:

$$n(\mathbf{r},t) = \int f(\mathbf{c}, \mathbf{r}, t)d\mathbf{c}.$$

For an arbitrary function $\phi(\mathbf{c})$ of the molecular velocity $\mathbf{c}$, $\langle \phi \rangle$ defined by

$$\langle \phi \rangle \equiv \frac{1}{n} \int \phi f d\mathbf{c}$$

is the average value of $\phi(\mathbf{c})$ at time t and position $\mathbf{r}$. In particular, taking $\mathbf{c}$ itself as ϕ, one obtains the macroscopic *flow velocity*,

$$\langle \mathbf{c} \rangle \equiv \mathbf{v} = \frac{1}{n} \int \mathbf{c} f d\mathbf{c}.$$

Subtraction of the flow velocity from the molecular velocity,

$$\mathbf{C} = \mathbf{c} - \mathbf{v},$$

is called the *peculiar velocity*: 'peculiar' means the balance after subtracting the 'common' part. The temperature T at a given time and space is defined through

$$\tfrac{1}{2} m \langle C^2 \rangle = \tfrac{3}{2} kT,$$

where m is the mass of a molecule.

The energy associated with translational motion of molecules is equal to the summation between the average energy of peculiar velocities and the energy of the macroscopic flow:

$$\tfrac{1}{2} m \langle (\mathbf{C} + \mathbf{v})^2 \rangle = \tfrac{1}{2} m \langle C^2 \rangle + \tfrac{1}{2} m v^2 .$$

When the molecules do not have rotational or vibrational degrees of freedom – we consider only such a case for the time being – the energy due to the peculiar velocities $mC^2/2$ is thus the thermal energy of the molecules. The flux of thermal energy may be decomposed as

$$n \langle \tfrac{1}{2} m C^2 \mathbf{c} \rangle = n \langle \tfrac{1}{2} m C^2 \mathbf{C} \rangle + n \langle \tfrac{1}{2} m C^2 \rangle \mathbf{v}.$$

The second term on the right-hand side represents heat transfer due to convection. The first term is the vector describing thermal conduction, which we set as

$$\mathbf{q} \equiv n \langle \tfrac{1}{2} m C^2 \mathbf{C} \rangle .$$

When deviation from equilibrium is not very great, the direction of this vector $\mathbf{q}$ agrees with that of decrease in the temperature T:

$$\mathbf{q} = -\kappa \frac{\partial T}{\partial \mathbf{r}}, \quad \frac{\partial}{\partial \mathbf{r}} \equiv \left(\frac{\partial}{\partial x}, \frac{\partial}{\partial y}, \frac{\partial}{\partial z} \right).$$

This coefficient of proportionality κ is the thermal conductivity.

For a uniform ideal gas, the pressure P is given by

$$P = nm \langle C_x^2 \rangle = nm \langle C_y^2 \rangle = nm \langle C_z^2 \rangle .$$

When the gas is not uniform, one extends this relation and considers the tensor,

$$\begin{pmatrix} P_{xx} & P_{xy} & P_{xz} \\ P_{yx} & P_{yy} & P_{yz} \\ P_{zx} & P_{zy} & P_{zz} \end{pmatrix} = nm \begin{pmatrix} \langle C_x^2 \rangle & \langle C_x C_y \rangle & \langle C_x C_z \rangle \\ \langle C_y C_x \rangle & \langle C_y^2 \rangle & \langle C_y C_z \rangle \\ \langle C_z C_x \rangle & \langle C_z C_y \rangle & \langle C_z^2 \rangle \end{pmatrix}.$$

P_{xx} represents the vertical pressure acting on a surface perpendicular to the x axis co-moving with the flow velocity; other diagonal components have similar meaning. The average of the diagonal elements,

$$P \equiv \tfrac{1}{3}(P_{xx} + P_{yy} + P_{zz}) = \tfrac{1}{3}nm \langle C^2 \rangle ,$$

is the hydrostatic pressure which is related to the temperature T via $P = nkT$.

When a gas with a uniform temperature moves with a flow velocity v_x parallel to the x axis and when the v_x is a function of z only, one obviously has, for reasons of symmetry,

$$P_{xy} = P_{yx} = 0, \quad P_{yz} = P_{zy} = 0.$$

Since

$$P_{xz} = nm \langle C_x C_z \rangle = nm \langle (c_x - v_x) C_z \rangle = n \langle mc_x C_z \rangle ,$$

$P_{xz} = P_{zx}$ describes the transfer rate of the x component of momentum in the z direction.

Difference between the pressure tensor and the hydrostatic pressure P corresponds to existence of viscosity in the gas; when the velocity gradient $\partial v_x/\partial z$ is not excessively large, we may set

$$P_{xz} = P_{zx} = -\eta \frac{\partial v_x}{\partial z}$$

with *viscosity* η as the coefficient of proportionality. It is clear that differences between the diagonal elements, P_{xx}, P_{yy}, P_{zz}, and the hydrostatic pressure P are infinitesimal quantities of order $(\partial v_x/\partial z)^2$, since these are even functions of $\partial v_x/\partial z$.

10.2 The Boltzmann equation

The integro-differential equation for the velocity distribution function $f(\mathbf{c}, \mathbf{r}, t)$ of the gas molecules is the Boltzmann equation, which we shall explain in this section.

The number of molecules found in $d\mathbf{r}$ at the position $\mathbf{r}$ and in $d\mathbf{c}$ around the velocity $\mathbf{c}$ at time t is

$$f(\mathbf{c}, \mathbf{r}, t)\, d\mathbf{c} d\mathbf{r}. \tag{10.1}$$

After a time interval dt, if no collisions are involved between molecules, exactly the same group of molecules are found in $d\mathbf{r}$ at the position $\mathbf{r} + \mathbf{c}dt$. Thus, without collisions between molecules and in the absence of external forces,

$$f(\mathbf{c}, \mathbf{r} + \mathbf{c}dt, t + dt)\, d\mathbf{c} d\mathbf{r} \tag{10.2}$$

is equal to (10.1).

In reality, owing to collisions between molecules, some of the molecules in the former domain (10.1) may not be found in the latter domain (10.2). Conversely, some of the molecules which did not belong to the former domain may enter the latter domain. Since the balance is proportional to $d\mathbf{c}d\mathbf{r}dt$, we may write it as

$(\partial_e f/\partial t)d\mathbf{c}d\mathbf{r}dt$ (the subscript e derives from 'encounter'); we thus have

$$f(\mathbf{c}, \mathbf{r} + \mathbf{c}dt, t + dt) - f(\mathbf{c}, \mathbf{r}, t) = \frac{\partial_e f}{\partial t} dt,$$

or

$$\frac{\partial f}{\partial t} + \mathbf{c} \cdot \frac{\partial f}{\partial \mathbf{r}} = \frac{\partial_e f}{\partial t}. \tag{10.3}$$

Here $\mathbf{c} \cdot \partial f/\partial \mathbf{r}$ means the scalar product, $c_x \partial f/\partial x + c_y \partial f/\partial y + c_z \partial f/\partial z$.

We now calculate

$$\frac{\partial_e f}{\partial t} d\mathbf{c}d\mathbf{r}dt \tag{10.4}$$

for spherically symmetric molecules such as rare-gas atoms. As stated above, this quantity is the balance between the number of molecules entering the domain $d\mathbf{c}d\mathbf{r}$ due to collisions during dt and that outgoing from the domain. First let us consider the latter rate. Assuming that a molecule with velocity $\mathbf{c}$ and one with $\mathbf{c}_1$ collide with each other resulting in velocities $\mathbf{c}'$ and $\mathbf{c}_1'$, respectively, we pay attention to relative velocities before and after the collision, $\mathbf{g} = \mathbf{c} - \mathbf{c}_1$ and $\mathbf{g}' = \mathbf{c}' - \mathbf{c}_1'$. Let θ be the angle between $\mathbf{g}'$ and $\mathbf{g}$; φ, the azimuthal angle of $\mathbf{g}'$ with respect to the $\mathbf{g}$ axis (see Figure 10.1); and $I(g, \theta)do$, the differential cross section of $\mathbf{g}'$ being deflected into the solid angle $\sin\theta d\theta d\varphi \equiv do$, where $g = |\mathbf{g}| = |\mathbf{g}'|$. Then the number of molecules outgoing from $d\mathbf{c}d\mathbf{r}$ and deflected into do during dt through collision with molecules in the velocity domain $d\mathbf{c}_1$ around $\mathbf{c}_1$ is

$$f(\mathbf{c})f(\mathbf{c}_1)gI(g,\theta)do\,d\mathbf{c}_1 d\mathbf{c}d\mathbf{r}dt, \tag{10.5}$$

where we have used an abbreviated notation $f(\mathbf{c})$ in place of $f(\mathbf{c}, \mathbf{r}, t)$.

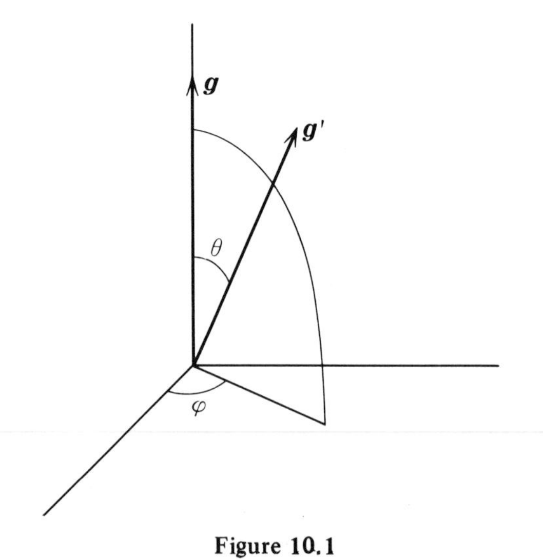

Figure 10.1

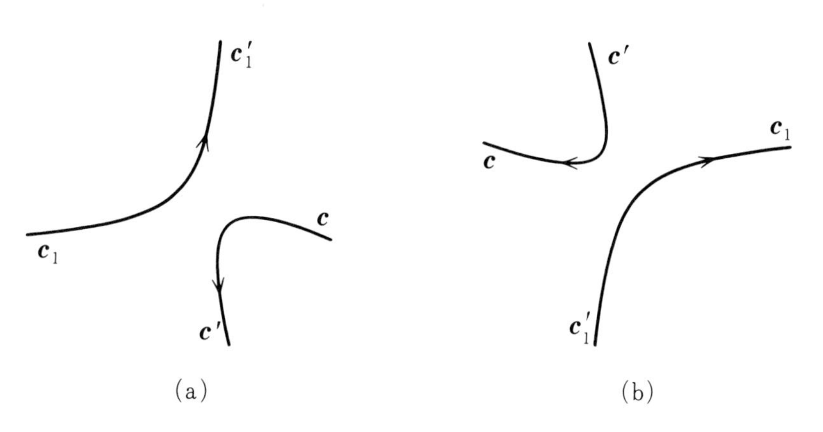

Figure 10.2 A collisional process (a) and its inverse process (b)

Next, those molecules entering $d\mathbf{c}d\mathbf{r}$ by collision must reach the velocity range $d\mathbf{c}$ around $\mathbf{c}$ after collision. For each process in which a molecule with velocity $\mathbf{c}$ collides with a molecule with velocity $\mathbf{c}_1$, their relative velocity deflected by θ and the velocities turning into $\mathbf{c}'$ and $\mathbf{c}_1'$, there exists an *inverse encounter* in which the velocities before collision are $\mathbf{c}'$ and $\mathbf{c}_1'$, the deflection angle θ, and the velocities after collision $\mathbf{c}$ and $\mathbf{c}_1$ (Figure 10.2). Corresponding to (10.5), the number of molecules entering into $d\mathbf{c}d\mathbf{r}$ is thus

$$f(\mathbf{c}')f(\mathbf{c}_1')gI(g,\theta)dod\mathbf{c}_1'd\mathbf{c}'\,d\mathbf{r}dt, \tag{10.6}$$

where $d\mathbf{c}_1'd\mathbf{c}'$ means the velocity domain that turns into $d\mathbf{c}_1d\mathbf{c}$ after collision.

Let us note that $d\mathbf{c}_1'd\mathbf{c}'$ may be replaced by $d\mathbf{c}_1d\mathbf{c}$. After doing so, we may subtract (10.5) from (10.6) and carry out integrations with respect to θ, φ, and the velocity $\mathbf{c}_1$ of the other molecule; the result is (10.4). We thus determine

$$\frac{\partial_e f}{\partial t} = \iint [f(\mathbf{c}')f(\mathbf{c}_1') - f(\mathbf{c})f(\mathbf{c}_1)]\, gI(g,\theta)dod\mathbf{c}_1$$

($do = \sin\theta d\theta d\varphi$). In the following we adopt the abbreviated notation,

$$\frac{\partial_e f}{\partial t} = \iint (f'f_1' - ff_1)gd\sigma d\mathbf{c}_1 \tag{10.7}$$

where

$$f_1 \equiv f(\mathbf{c}_1,\mathbf{r},t), \quad f' \equiv f(\mathbf{c}',\mathbf{r},t), \quad f_1' \equiv f(\mathbf{c}_1',\mathbf{r},t), \quad d\sigma \equiv I(g,\theta)do. \tag{10.8}$$

10.3 Boltzmann's *H* theorem and the Maxwellian velocity distribution

From the Boltzmann equation, the *H* theorem and the Maxwellian velocity distribution are derived, which we shall describe in the following.

For the cases that the velocity distribution function is independent of the position $\mathbf{r}$, the Boltzmann equation takes the form,

$$\frac{\partial f}{\partial t} = \iint (f'f_1' - ff_1) g d\sigma d\mathbf{c}_1,$$

where $f \equiv f(\mathbf{c}, t), f_1 \equiv f(\mathbf{c}_1, t), f' \equiv f(\mathbf{c}', t), f_1' \equiv f(\mathbf{c}_1', t)$. Defining the quantity H by

$$H \equiv \int f \ln f d\mathbf{c}, \tag{10.9}$$

we have

$$\frac{dH}{dt} = \int (\ln f + 1) \frac{\partial f}{\partial t} d\mathbf{c} = \iiint (\ln f + 1)(f'f_1' - ff_1) g d\sigma d\mathbf{c}_1 d\mathbf{c}.$$

Since the integral is invariant under interchange between $\mathbf{c}$ and $\mathbf{c}_1$ in the integrand, $\ln f + 1$ may be replaced by $(\ln f + \ln f_1 + 2)/2$. We then add an integral representing the contribution of inverse collisions, and divide the result by two; replacing $d\mathbf{c}_1' \, d\mathbf{c}'$ by $d\mathbf{c}_1 d\mathbf{c}$, we obtain

$$\frac{dH}{dt} = -\frac{1}{4} \iiint (\ln f'f_1' - \ln ff_1)(f'f_1' - ff_1) g d\sigma d\mathbf{c}_1 d\mathbf{c}.$$

Since the signs of $\ln f'f_1' - \ln ff_1'$ and $f'f_1' - ff_1$ are the same, the integrand is a non-negative function, that is,

$$\frac{dH}{dt} \leqslant 0. \tag{10.10}$$

This is called Boltzmann's H Theorem. Since $-H$ corresponds to entropy per unit volume, the H theorem states that the entropy of a gas generally increases as a consequence of molecular collisions.

A stationary state $dH/dt = 0$ is achieved when the 'condition of detailed balancing',

$$f'f_1' = ff_1,$$

is satisfied for all the values of $\mathbf{c}$ and $\mathbf{c}_1$, that is,

$$\ln f(\mathbf{c}') + \ln f(\mathbf{c}_1') = \ln f(\mathbf{c}) + \ln f(\mathbf{c}_1). \tag{10.11}$$

In a stationary state the sum of $\ln f$ for the two molecules thus remains unchanged in the event of collision. Since the total values of the momentum $m\mathbf{c}$ and the energy $mc^2/2$ also remain unchanged in the event of collision, $\ln f$ may be taken as a linear combination of those quantities,

$$\ln f(\mathbf{c}) = \alpha^{(1)} + \boldsymbol{\alpha}^{(2)} \cdot m\mathbf{c} - \alpha^{(3)} mc^2/2;$$

the relation (10.11) is then satisfied. No forms other than that make the sum invariant. (For, in the event of collision between two spherically symmetric molecules, six relations arising from the two angular variables θ and φ, three representing conservation of the three components of momentum, and one of energy conservation completely determine the six components of the velocities $\mathbf{c}'$ and $\mathbf{c}_1'$ after collision.)

The foregoing formula for $\ln f$ may also be written in the form:

$$f(\mathbf{c}) = \alpha^{(0)} \exp[-\alpha^{(3)} m \mid \mathbf{c} - \boldsymbol{\alpha} \mid^2 /2].$$

The constants $\alpha^{(0)}$, $\alpha^{(3)}$, and $\boldsymbol{\alpha}$ appearing in this formula may be interpreted in the following way.

The number n of molecules in a unit volume is expressed as

$$n = \int f(\mathbf{c}) d\mathbf{c} = \alpha^{(0)} \left(\frac{2\pi}{m\alpha^{(3)}} \right)^{3/2}.$$

Likewise the flow velocity $\mathbf{v}$ may be written as

$$n\mathbf{v} = \int \mathbf{c} f(\mathbf{c}) d\mathbf{c} = \int \boldsymbol{\alpha} f d\mathbf{c} + \int (\mathbf{c} - \boldsymbol{\alpha}) f d(\mathbf{c} - \boldsymbol{\alpha}).$$

The first term on the right-hand side is equal to $n\boldsymbol{\alpha}$; the second term vanishes since the integrand is an odd function of $\mathbf{c} - \boldsymbol{\alpha}$. Thus $\boldsymbol{\alpha}$ is equal to $\mathbf{v}$. Finally we use

$$\tfrac{3}{2} kT = \tfrac{1}{2} m \langle C^2 \rangle, \quad \mathbf{C} = \mathbf{c} - \mathbf{v},$$

to obtain

$$\frac{m}{2n} \int C^2 f d\mathbf{c} = \frac{m}{2} \left(\frac{m\alpha^{(3)}}{2\pi} \right)^{3/2} \int_0^\infty C^2 \exp(-\alpha^{(3)} \tfrac{1}{2} m C^2) 4\pi C^2 dC = \frac{3}{2\alpha^{(3)}}.$$

Thus $\alpha^{(3)} = 1/kT$. We have consequently derived the Maxwellian velocity distribution:

$$f(\mathbf{c}) = n \left(\frac{m}{2\pi kT} \right)^{3/2} \exp(-m \mid \mathbf{c} - \mathbf{v} \mid^2 /2kT). \tag{10.12}$$

10.4 Velocity distribution not far from the Maxwellian

Since we treat transport phenomena in gasses with small temperature or velocity gradients, the velocity distribution $f(\mathbf{c}, \mathbf{r}, t)$ of the molecules may be regarded as close to the Maxwellian $f^{(0)}$. Naturally, the $f^{(0)}$ must appropriately describe the number density n, the flow velocity $\mathbf{v}$, and the temperature T at $\mathbf{r}$ and t, i.e.

$$\int f^{(0)} d\mathbf{c} = \int f d\mathbf{c} = n,$$

$$\frac{1}{n} \int f^{(0)} \mathbf{c} d\mathbf{c} = \frac{1}{n} \int f \mathbf{c} d\mathbf{c} = \mathbf{v},$$

$$\frac{1}{n} \int f^{(0)} \tfrac{1}{2} m C^2 d\mathbf{c} = \frac{1}{n} \int f \tfrac{1}{2} m C^2 d\mathbf{c} = \tfrac{3}{2} kT.$$

Setting

$$f = f^{(0)} (1 + \Phi), \tag{10.13}$$

we thus have

$$\int f^{(0)}\Phi d\mathbf{c} = 0, \quad \int f^{(0)}\Phi \mathbf{C} d\mathbf{c} = 0, \quad \int f^{(0)}\Phi C^2 d\mathbf{c} = 0. \tag{10.14}$$

For example, each element of the tensor

$$\Phi = \frac{m}{2kT}(\mathbf{CC} - \tfrac{1}{3}C^2\mathbf{1}) \times \left(\text{an arbitrary function of } \frac{mC^2}{2kT}\right)$$

constructed out of the peculiar velocity $\mathbf{C}$ satisfies the three conditions stated above.

When one deals with a probability distribution close to the normal distribution in ordinary statistics, the Hermite polynomials are useful. Analogously for a three-dimensional velocity distribution, one uses Sonine's polynomials (D. Burnett 1935). Sonine's polynomials $S_m^{(n)}(x)$ for any real numbers x and m are defined via

$$(1-s)^{-m-1}\exp\left(-\frac{xs}{1-s}\right) = \sum_{n=0}^{\infty} S_m^{(n)}(x)s^n,$$

where $0 < s < 1$. In particular it is easy to see

$$S_m^{(0)} = 1, \quad S_m^{(1)}(x) = m + 1 - x. \tag{10.15}$$

Since

$$\begin{aligned}
&(1-s)^{-m-1}(1-t)^{-m-1}\int_0^\infty \exp\left[-x\left(1+\frac{s}{1-s}+\frac{t}{1-t}\right)\right]x^m dx \\
&= (1-s)^{-m-1}(1-t)^{-m-1}\int_0^\infty \exp\left[-\frac{x(1-st)}{(1-s)(1-t)}\right]x^m dx \\
&= (1-st)^{-m-1}\Gamma(m+1),
\end{aligned}$$

we obtain the following orthogonality condition:

$$\int_0^\infty e^{-x}S_m^{(p)}(x)S_m^{(q)}(x)x^m dx = \frac{\Gamma(m+p+1)}{p!}\delta_{pq}, \tag{10.16}$$

from comparison of the coefficients of $s^p t^q$. (For $p = q, \delta_{pq} = 1$; for $p \neq q$, $\delta_{pq} = 0$.)

Generally for two arbitrary functions, ϕ and ψ, of the molecular velocities, we define their inner product by

$$(\phi,\psi) \equiv \frac{1}{n}\int f^{(0)}\phi \cdot \psi d\mathbf{c}.$$

When ϕ and ψ are vectors, $\phi \cdot \psi$ here means a scalar product; when both are tensors, it represents the scalar product between the tensors. For example, for

$$\psi^{(r)} \equiv \left(\frac{m}{2kT}\right)^{1/2}\mathbf{C}S_{3/2}^{(r)}\left(\frac{mC^2}{2kT}\right), \quad r = 0, 1, \ldots, \tag{10.17}$$

we have

$$(\psi^{(r)},\psi^{(s)}) = \frac{1}{n}\int_0^\infty f^{(0)}\frac{mC^2}{2kT}S_{3/2}^{(r)}\left(\frac{mC^2}{2kT}\right)S_{3/2}^{(s)}\left(\frac{mC^2}{2kT}\right)$$
$$\times 4\pi C^2 dC = \frac{2}{\sqrt{\pi}}\frac{1}{r!}\Gamma\left(r+\frac{5}{2}\right)\delta_{rs}. \qquad (10.18)$$

Also for

$$\varphi^{(r)} \equiv \frac{m}{2kT}(\mathbf{CC}-\tfrac{1}{3}C^2\mathbf{1})S_{5/2}^{(r)}\left(\frac{mC^2}{2kT}\right), \quad r = 0, 1, \ldots, \qquad (10.19)$$

we have, with the aid of $\mathbf{CC}:\mathbf{CC} = C^4$, $\mathbf{1}:\mathbf{CC} = C^2$, and $\mathbf{1}:\mathbf{1} = 3$,

$$(\varphi^{(r)}, \varphi^{(s)}) = \frac{4}{3}\frac{1}{\sqrt{\pi}}\frac{1}{r!}\Gamma\left(r+\frac{7}{2}\right)\delta_{rs}. \qquad (10.20)$$

In this way, a set of orthogonal functions of the molecular velocity can be constructed. $\psi^{(r)}$ and $\varphi^{(r)}$ will be needed when we deal with thermal conductivity and viscosity, respectively. A linear combination of the $\varphi^{(r)}$ and a linear combination of the $\psi^{(r)}$ without $\psi^{(0)}$ may be used for Φ in (10.13), since they satisfy the condition (10.14).

Now, substituting $f = f^{(0)}(1+\Phi)$ in Boltzmann's collision term (10.7) and disregarding Φ^2 we have

$$\frac{\partial_e f}{\partial t} = f^{(0)}\iint f_1^{(0)}(\Phi' + \Phi_1' - \Phi - \Phi_1)g\,d\sigma\,d\mathbf{c}_1,$$

where Φ_1, Φ', etc. are shorthand notation similar to f_1, f', etc.

We here define a linear operator J acting on an arbitrary function ϕ of the molecular velocity $\mathbf{c}$ as follows:

$$J\phi \equiv \frac{1}{n}\iint f_1^{(0)}(\phi + \phi_1 - \phi' - \phi_1')g\,d\sigma\,d\mathbf{c}_1. \qquad (10.21)$$

Boltzmann's collision term is then expressed in a concise form,

$$\frac{\partial_e f}{\partial t} = -nf^{(0)}J\Phi, \quad f = f^{(0)}(1+\Phi). \qquad (10.22)$$

For two arbitrary functions, ϕ and ψ, of $\mathbf{c}$, the inner product between ϕ and $J\psi$,

$$(\phi, J\psi) = \frac{1}{n^2}\iiint f^{(0)}f_1^{(0)}\phi\cdot(\psi + \psi_1 - \psi' - \psi_1')g\,d\sigma\,d\mathbf{c}_1\,d\mathbf{c},$$

may be transformed into

$$(\phi, J\psi) = \frac{1}{4n^2}\iiint f^{(0)}f_1^{(0)}(\phi + \phi_1 - \phi' - \phi_1')\cdot(\psi + \psi_1 - \psi' - \psi_1')g\,d\sigma\,d\mathbf{c}_1\,d\mathbf{c}, \qquad (10.23)$$

in a way analogous to that of Section 10.3. Hence,

$$(\phi, J\psi) = (J\phi, \psi). \tag{10.24}$$

That is, J is a symmetric operator. Furthermore,

$$(\phi, J\phi) \geqslant 0. \tag{10.25}$$

That is, J is positive definite.

We now single out

$$\varphi^{(0)} = \frac{m}{2kT}(\mathbf{CC} - \tfrac{1}{3}C^2\mathbf{1})$$

from among the $\varphi^{(r)}$ defined in (10.19), and calculate

$$(\varphi^{(0)}, J\varphi^{(0)}) = \frac{1}{4}\frac{1}{n^2}\iiint f^{(0)} f_1^{(0)} \left(\frac{m}{2kT}\right)^2 \times (\mathbf{CC} + \mathbf{C}_1\mathbf{C}_1 - \mathbf{C}'\mathbf{C}' - \mathbf{C}_1'\mathbf{C}_1')^2\, g\,d\sigma d\mathbf{c}_1 d\mathbf{c},$$

where $(\mathbf{CC} + \ldots)^2$ in the integrand implies a scalar product. In terms of the centre-of-mass velocity $\mathbf{G} = (\mathbf{C} + \mathbf{C}_1)/2$, and the relative velocities, $\mathbf{g} = \mathbf{C} - \mathbf{C}_1$ and $\mathbf{g}' = \mathbf{C}' - \mathbf{C}_1'$, before and after collision, we express

$$\mathbf{CC} + \mathbf{C}_1\mathbf{C}_1 - \mathbf{C}'\mathbf{C}' - \mathbf{C}_1'\mathbf{C}_1' = (\mathbf{gg} - \mathbf{g}'\mathbf{g}')/2.$$

With the aid of

$$(\mathbf{gg} - \mathbf{g}'\mathbf{g}') : (\mathbf{gg} - \mathbf{g}'\mathbf{g}') = 2g^4(1 - \cos^2\theta),$$

$$\exp\left[-\frac{m}{2kT}(C^2 + C_1^2)\right] = \exp\left[-\frac{m}{kT}G^2 - \frac{m}{4kT}g^2\right],$$

$$d\mathbf{c}_1 d\mathbf{c} = d\mathbf{C}_1 d\mathbf{C} = d\mathbf{G}d\mathbf{g},$$

we then determine

$$(\varphi^{(0)}, J\varphi^{(0)}) = 4\Omega^{(2,2)}. \tag{10.26}$$

Here the 'Ω integrals' are generally defined as

$$\Omega^{(l,r)} = \left(\frac{kT}{2\pi m^*}\right)^{1/2} \int_0^\infty \exp(-g^{*2}) g^{*2r+3} \int (1 - \cos^l\theta)\, d\sigma dg^*, \tag{10.27}$$

$$g^* = (m^*/2kT)^{1/2} g, \quad l = 1, 2, \ldots, \quad r = l, l+1, \ldots \tag{10.28}$$

where m^* is the reduced mass; in this particular case, it is $m/2$.

In this way we obtain the formulas in Table 10.1. Similar calculations may be carried out for $\psi^{(r)}$ defined in (10.17); the results are also listed in the same table ($J\psi^{(0)} \equiv 0$).

The integrals $\Omega^{(l,r)}$ represent various mean values of the collisional effects between molecules, and are generally functions of temperature T. They satisfy a

Table 10.1

$(\varphi^{(0)}, J\varphi^{(0)}) = 4\Omega^{(2,2)}$
$(\varphi^{(1)}, J\varphi^{(0)}) = 7\Omega^{(2,2)} - 2\Omega^{(2,3)}$
$(\varphi^{(1)}, J\varphi^{(1)}) = (301/12)\Omega^{(2,2)} - 7\Omega^{(2,3)} + \Omega^{(2,4)}$
$(\psi^{(1)}, J\psi^{(1)}) = 4\Omega^{(2,2)}$
$(\psi^{(2)}, J\psi^{(1)}) = 7\Omega^{(2,2)} - 2\Omega^{(2,3)}$
$(\psi^{(2)}, J\psi^{(2)}) = (77/4)\Omega^{(2,3)} + \Omega^{(2,4)}$

S. Chapman and T. G. Cowling (1970), *The Mathematical Theory of Non-Uniform Gases*, 3rd ed., Cambridge Univ. Press.

recursive formula,

$$T\frac{d}{dT}\Omega^{(l,r)} = \Omega^{(l,r+1)} - \left(r + \frac{3}{2}\right)\Omega^{(l,r)}. \tag{10.29}$$

When interaction between molecules may be expressed by an inverse-power potential,

$$U(r) = \lambda r^{-p}, \qquad \lambda \gtrless 0, p > 1,$$

the Ω integrals are proportional to the $(1/2 - 2/p)$th power of the temperature T. This may be understood from the fact that the quantity with the dimension of area constructed out of p, λ, and kT is $(\lambda/kT)^{2/p}$. In particular when $p = 4$, the Ω integrals are independent of the temperature. Such a case is called Maxwell's model.

In Maxwell's model, that is, when the intermolecular potential $U(r)$ is proportional to r^{-4}, both sides of (10.29) as well as $(\varphi^{(1)}, J\varphi^{(0)})$ etc. all vanish. In fact, in this model the $\varphi^{(r)}$ and $\psi^{(r)}$ are eigenfunctions of J; hence for all $r \neq s$, $(\varphi^{(r)}, J\varphi^{(s)}) = 0$ and $(\psi^{(r)}, J\psi^{(s)}) = 0$.

10.5 Expressions for viscosity

Consider a case in which a gas with uniform temperature and density moves with a flow velocity v_x parallel to the x axis and v_x is a function of z alone. In these circumstances, since the velocity distribution function f is independent of x and y, the Boltzmann equation reads

$$\frac{\partial f}{\partial t} + c_z\frac{\partial f}{\partial z} = \iint (f'f_1' - ff_1)g\,d\sigma\,d\mathbf{c}_1. \tag{10.30}$$

If v_x is constant, the left-hand side of (10.30) would vanish; the molecular velocities take the Maxwellian distribution

$$f^{(0)} = n\left(\frac{m}{2\pi kT}\right)^{3/2} \exp\left(-\frac{mC^2}{2kT}\right), \qquad \mathbf{C} = \mathbf{c} - \mathbf{v}. \tag{10.31}$$

When the rate of change $\partial v_x/\partial z$ is not large, one expects that the velocity distribution function f does not differ substantially from this $f^{(0)}$.

Approximate values of the left-hand side of (10.30) may be obtained by substituting $f^{(0)}$ in place of f. Increase in temperature owing to viscous friction can be neglected in these circumstances so that $\partial f/\partial t = 0$;

$$c_z \frac{\partial f^{(0)}}{\partial z} = -C_z \frac{\partial f^{(0)}}{\partial C_x} \frac{\partial v_x}{\partial z} = f^{(0)} \frac{mC_x C_z}{kT} \frac{\partial v_x}{\partial z}.$$

In the right-hand side of (10.30), we may substitute the expression

$$f = f^{(0)}(1 + \Phi), \qquad \Phi = \Phi(\mathbf{C}), \tag{10.32}$$

and disregard Φ^2. Then (10.30) reduces to

$$\frac{mC_x C_z}{kT} \frac{\partial v_x}{\partial z} = -nJ\Phi \tag{10.33}$$

$$J\Phi = \frac{1}{n} \int\!\!\int f_1^{(0)}(\Phi + \Phi_1 - \Phi' - \Phi_1')g d\sigma d\mathbf{c}_1 \tag{10.34}$$

where J is the integral operator defined in (10.21).

We here set

$$\Phi = -\frac{1}{n} \frac{mC_x C_z}{2kT} B \frac{\partial v_x}{\partial z} \tag{10.35}$$

where B is looked upon as a function of $mC^2/2kT$. Substitution of this in (10.33) yields

$$J\left(\frac{mC_x C_z}{2kT} B\right) = 2 \frac{mC_x C_z}{2kT}. \tag{10.36}$$

The xz component of the pressure tensor is calculated as

$$P_{xz} = P_{zx} = m \int f C_x C_z d\mathbf{c} = m \int f^{(0)} \Phi C_x C_z d\mathbf{c}$$

$$= -\frac{\partial v_x}{\partial z} \frac{2kT}{n} \int f^{(0)} \left(\frac{mC_x C_z}{2kT}\right)^2 B d\mathbf{c}.$$

The viscosity η defined by $P_{xz} = -\eta \partial v_x/\partial z$ is thus given by the following formula:

$$\eta = \frac{2kT}{n} \int f^{(0)} \left(\frac{mC_x C_z}{2kT}\right)^2 B d\mathbf{c}.$$

Average of $C_x^2 C_z^2$ over the directions of $\mathbf{C}$ yields $C^4/15$, so that

$$\eta = \frac{2kT}{15n} \int f^{(0)} \left(\frac{mC^2}{2kT}\right)^2 B d\mathbf{c}. \tag{10.37}$$

We now introduce the set of orthogonal functions

$$\varphi^{(r)} = \frac{m}{2kT}(\mathbf{CC} - \tfrac{1}{3}C^2 \mathbf{1}) S_{5/2}^{(r)}\left(\frac{mC^2}{2kT}\right), \tag{10.38}$$

the properties of which have already been investigated; we thereby expand

$$\frac{m}{2kT}(\mathbf{CC}-\tfrac{1}{3}C^2\mathbf{1})B=\sum_{r=0}^{\infty}b_r\varphi^{(r)}. \tag{10.39}$$

Since (10.36) is the xz component of

$$J\left\{\frac{m}{2kT}(\mathbf{CC}-\tfrac{1}{3}C^2\mathbf{1})B\right\}=2\frac{m}{2kT}(\mathbf{CC}-\tfrac{1}{3}C^2\mathbf{1}),$$

we find

$$\sum_{r=0}^{\infty}b_rJ\varphi^{(r)}=2\varphi^{(0)}. \tag{10.40}$$

On account of

$$(\mathbf{CC}-\tfrac{1}{3}C^2\mathbf{1}):(\mathbf{CC}-\tfrac{1}{3}C^2\mathbf{1})=\tfrac{2}{3}C^4,$$

the expression (10.37) for η may be transformed into

$$\eta=\tfrac{1}{5}kT(\varphi^{(0)},\ \sum_{r=0}^{\infty}b_r\varphi^{(r)})=\tfrac{1}{2}kTb_0. \tag{10.41}$$

Thus η is determined solely by b_0 among the coefficients in (10.39).

Carrying out scalar products of both sides of (10.40) by $\varphi^{(s)}$ and integrating the result, we may determine b_0 as a solution to the simultaneous equations,

$$\begin{aligned}\sum_{r=0}^{\infty}b_r[\varphi^{(s)},\varphi^{(r)}]&=2(\varphi^{(0)},\varphi^{(0)})=5 \qquad (s=0)\\ &=0 \qquad\qquad\qquad\qquad (s\neq 0),\end{aligned} \tag{10.42}$$

where we have set

$$(\varphi^{(s)},J\varphi^{(r)})\equiv[\varphi^{(r)},\varphi^{(s)}]=[\varphi^{(s)},\varphi^{(r)}].$$

For different r and s, $[\varphi^{(r)},\ \varphi^{(s)}]$ vanish in Maxwell's model, i.e. for the inverse fourth-power potential; generally they take on small values. Hence as a first approximation to b_0, we may set all of the $[\varphi^{(r)},\varphi^{(s)}]$ for $r\neq s$ equal to zero and take the solution of

$$b_0[\varphi^{(0)},\varphi^{(0)}]=5.$$

With the aid of (10.26) we thus obtain the first approximation to the viscosity η as follows:

$$[\eta]_1=\frac{5kT}{8\Omega^{(2,2)}}. \tag{10.43}$$

For the inverse fourth-power potential, this is exactly equal to the accurate value.

The second approximation to b_0 is obtained from the two by two terms of

(10.42),

$$b_0[\varphi^{(0)},\varphi^{(0)}]+b_1[\varphi^{(0)},\varphi^{(1)}]=5$$
$$b_0[\varphi^{(1)},\varphi^{(0)}]+b_1[\varphi^{(1)},\varphi^{(1)}]=0. \tag{10.44}$$

Alternatively we note the expression

$$b_0=\frac{5}{[\varphi^{(0)},\varphi^{(0)}]}\left\{1+\frac{[\varphi^{(0)},\varphi^{(1)}]^2}{[\varphi^{(0)},\varphi^{(0)}][\varphi^{(1)},\varphi^{(1)}]}\right\}$$

retaining up to the square of a small quantity, $[\varphi^{(0)}, \varphi^{(1)}]$. With the aid of $[\varphi^{(0)}, \varphi^{(1)}]$ in Table 10.1 and the relation in the first approximation,

$$[\varphi^{(1)},\varphi^{(1)}]=(49/3)\Omega^{(2,2)},$$

we may adopt as the second approximation to η

$$[\eta]_2=[\eta]_1(1+\delta),\quad \delta=\frac{3}{49}\left(\frac{\Omega^{(2,3)}}{\Omega^{(2,2)}}-\frac{7}{2}\right)^2. \tag{10.45}$$

For an inverse pth-power potential $U(r)=\lambda r^{-p}$,

$$\delta=\frac{3}{49}\left(\frac{1}{2}-\frac{2}{p}\right)^2.$$

This number is relatively small; a sufficient accuracy is achieved generally by calculating up to $[\eta]_2$.

The expression (10.43) and the solution to (10.44) written in terms of the Ω integrals are called the first and the second approximations of Chapman and Cowling. The simplified expression (10.45) is sometimes called Kihara's second approximation.

10.6 Expressions for thermal conductivity

The method described in the previous section is equally applicable to thermal conductivity for a single-component gas constituting of molecules with no internal degrees of freedom.

For such a gas at rest, we consider the case that the temperature T is a function of the position. Since the pressure must be uniform, the number density n of the molecules is inversely proportional to T; thus n likewise is a function of the position.

If T is uniform, the molecular velocities would take the Maxwellian distribution,

$$f^{(0)}=n\left(\frac{m}{2\pi kT}\right)^{3/2}\exp\left(-\frac{mC^2}{2kT}\right). \tag{10.46}$$

If the spatial variation of T is not excessively large, one expects that the velocity distribution function f may not differ substantially from this $f^{(0)}$. We thus adopt

$f^{(0)}$ as the f on the left-hand side of the Boltzmann equation. In these circumstances the heat flow is nearly in a stationary state, so that we may set $\partial f/\partial t = 0$. On account of

$$\mathbf{c}\cdot\frac{\partial f^{(0)}}{\partial \mathbf{r}} = f^{(0)}\mathbf{c}\cdot\left(\frac{\partial \ln f^{(0)}}{\partial \ln T}\frac{\partial \ln T}{\partial \mathbf{r}} + \frac{\partial \ln f^{(0)}}{\partial \ln n}\frac{\partial \ln n}{\partial \mathbf{r}}\right)$$

$$= f^{(0)}\left(\frac{\partial \ln f^{(0)}}{\partial \ln T} - \frac{\partial \ln f^{(0)}}{\partial \ln n}\right)\mathbf{C}\cdot\frac{\partial \ln T}{\partial \mathbf{r}}$$

$$= f^{(0)}\left(\frac{mC^2}{2kT} - \frac{5}{2}\right)\mathbf{C}\cdot\frac{\partial \ln T}{\partial \mathbf{r}},$$

we may substitute $f = f^{(0)}(1 + \Phi)$ on the right-hand side of the Boltzmann equation and disregard Φ^2; the result is

$$\left(\frac{mC^2}{2kT} - \frac{5}{2}\right)\mathbf{C}\cdot\frac{\partial \ln T}{\partial \mathbf{r}} = -nJ\Phi, \tag{10.47}$$

where J takes the same form as in the previous section.

As for Φ, we set

$$\Phi = -\frac{1}{n}\mathbf{C}\cdot\frac{\partial \ln T}{\partial \mathbf{r}}A,$$

where A is a function of $mC^2/2kT$. Substitution of this in (10.47) yields

$$J(\mathbf{C}A) = \mathbf{C}\left(\frac{mC^2}{2kT} - \frac{5}{2}\right). \tag{10.48}$$

In addition, Φ must satisfy the conditions in (10.14); one of those conditions requires

$$\int f^{(0)}\mathbf{C}\mathbf{C}A d\mathbf{c} = 0 \tag{10.49}$$

and then the rest is automatically satisfied.

Since the flow $\mathbf{q}$ of thermal energy is

$$\mathbf{q} = \int f\frac{m}{2}C^2\mathbf{C}d\mathbf{c}$$

$$= \frac{m}{2}\int f^{(0)}\Phi C^2\mathbf{C}d\mathbf{c}$$

$$= -\frac{m}{2n}\int f^{(0)}C^2\mathbf{C}\mathbf{C}A d\mathbf{c}\cdot\frac{\partial \ln T}{\partial \mathbf{r}}$$

$$= -\frac{m}{6n}\int f^{(0)}C^4 A d\mathbf{c}\frac{\partial \ln T}{\partial \mathbf{r}},$$

the thermal conductivity κ defined through $\mathbf{q} = -\kappa\,\partial T/\partial \mathbf{r}$ is given by

$$\kappa = \frac{m}{6nT}\int f^{(0)} C^4 A dc.$$

Referring to (10.49), we may also transform it into the form,

$$\kappa = \frac{2k^2T}{3mn}\int f^{(0)}\frac{mC^2}{2kT}\left(\frac{mC^2}{2kT} - \frac{5}{2}\right)Adc. \tag{10.50}$$

We now introduce the set of orthogonal functions,

$$\psi^{(r)} = \left(\frac{m}{2kT}\right)^{1/2} \mathbf{C}S_{3/2}^{(r)}\left(\frac{mC^2}{2kT}\right), \quad r = 0, 1, \ldots,$$

the properties of which have already been investigated; we thereby expand

$$\left(\frac{m}{2kT}\right)^{1/2} \mathbf{C}A = -\sum_{r=0}^{\infty} a_r \psi^{(r)}.$$

On account of (10.49), $a_0 = 0$. Substitution of the foregoing formula in (10.48) yields

$$\sum_{r=1}^{\infty} a_r J\psi^{(r)} = \psi^{(1)}. \tag{10.51}$$

From (10.50) we have

$$\kappa = \frac{2k^2T}{3m}\left(\psi^{(1)}, \sum_{r=1}^{\infty} a_r\psi^{(r)}\right) = \frac{5k^2T}{2m}a_1, \tag{10.52}$$

so that κ is determined by the coefficient a_1.

Carrying out scalar products of both sides of (10.51) by $\psi^{(s)}$ ($s = 1, 2, \ldots$) and integrating the results, we obtain the simultaneous equations to determine a_1,

$$\begin{aligned}\sum_{r=1}^{\infty} a_r[\psi^{(s)}, \psi^{(r)}] &= (\psi^{(1)}, \psi^{(1)}) = 15/4 \quad (s = 1)\\ &= 0 \qquad\qquad\qquad\qquad (s \neq 1)\end{aligned} \tag{10.53}$$

where we have set

$$(\psi^{(s)}, J\psi^{(r)}) \equiv [\psi^{(s)}, \psi^{(r)}] = [\psi^{(r)}, \psi^{(s)}].$$

With the aid of the formulas in Table 10.1 and the approximate value $21\,\Omega^{(2,2)}/2$ for $[\psi^{(2)}, \psi^{(2)}]$, in a way quite similar to that of the previous section, the first and second approximations for the thermal conductivity κ are obtained as

$$[\kappa]_1 = \frac{3k}{2}\frac{25kT}{16m}\frac{1}{\Omega^{(2,2)}}, \tag{10.54}$$

$$[\kappa]_2 = [\kappa]_1(1 + \delta), \quad \delta = \frac{2}{21}\left(\frac{\Omega^{(2,3)}}{\Omega^{(2,2)}} - \frac{7}{2}\right)^2. \tag{10.55}$$

For an inverse fourth-potential, $[\kappa]_1 = [\kappa]_2$ is equal to the exact value.

As we have stated above, the thermal conductivity κ is determined solely by a_1 among the coefficients in the expansion $\Sigma a_r \psi^{(r)}$. In addition, that a_2, etc. are actually small can be seen from

$$\frac{a_2}{a_1} \doteqdot \frac{4}{21}\left(\frac{\Omega^{(2,3)}}{\Omega^{(2,2)}} - \frac{7}{2}\right).$$

This also indicates how the functional form of the molecular velocity distribution f may deviate from the Maxwellian in the presence of the temperature gradient. Thus in terms of $f^{(0)}$ in (10.46), we have

$$\begin{aligned} f &= f^{(0)}\left[1 + \frac{a_1}{n} S_{3/2}^{(1)}\left(\frac{mC^2}{2kT}\right) \mathbf{C} \cdot \frac{\partial \ln T}{\partial \mathbf{r}}\right] \\ &= f^{(0)}\left[1 + \frac{a_1}{n}\left(\frac{5}{2} - \frac{mC^2}{2kT}\right) \mathbf{C} \cdot \frac{\partial \ln T}{\partial \mathbf{r}}\right], \end{aligned} \tag{10.56}$$

where

$$a_1 = 15/16\Omega^{(2,2)}. \tag{10.57}$$

Figure 10.3 depicts $\iint f dC_x dC_y$ as a function of $(m/2kT)^{1/2} C_z$, where the positive z direction is chosen in the direction of increasing temperature. Naturally, fast molecules are directed more from the high temperature domain to the low temperature domain; they transport the thermal energy. Slow molecules are directed more from the low energy part to the high energy part; they act to cancel the overall flow of particles. Equations (10.56) and (10.57) will be used later when we treat thermal diffusion phenomena.

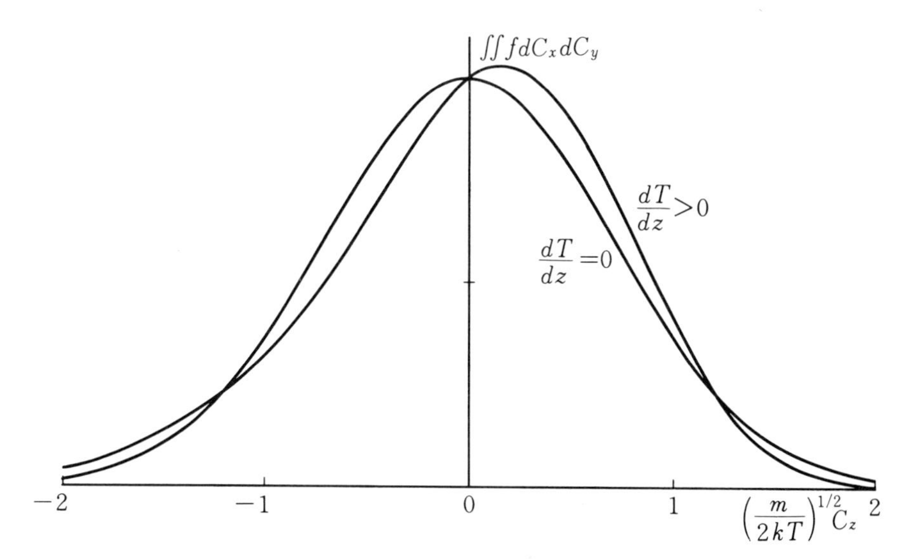

Figure 10.3

The expressions (10.54) and (10.55), for thermal conductivity, are applicable to those molecules without internal degrees of freedom such as the rare gases. In these cases, using $c_v = 3k/2$ for the constant-volume specific heat per molecule, we find, through comparison with the first approximation (10.43) of the viscosity, that the relation

$$[\kappa]_1 = \frac{5}{2}\frac{c_v}{m}[\eta]_1 \tag{10.58}$$

holds for gases of the same kind at the same temperature. Thus,

$$m\kappa/c_v\eta \equiv \mathrm{f} \tag{10.59}$$

must be close to a constant, 2.5. Measured values are very close to this number.

For hydrogen H_2, as the temperature is decreased below 0 °C, the ratio γ between the constant-pressure specific heat and the constant-volume specific heat increases and approaches the rare-gas value 5/3 at about −200 °C. This fact indicates that the molecular energy at such a low temperature consists mostly of that associated with translational motion. Hence one expects that below −200 °C the value of f approaches 2.5. The measured values by E. Eucken (1913) and J.B. Ubbink (1948) indeed substantiate this expectation.

In a gas of polyatomic molecules, a portion of thermal energy is transported by the internal motion of molecules such as molecular rotation. In this connection let us take up Eucken's idea which has been known for quite some time. For this purpose we first show that the value of f in (10.59) may be interpreted as the ratio between the mean free path l effective for thermal conduction and the mean free path l' effective for viscosity.

When the flow velocity v_x is a function of z alone, the viscosity η is defined by

$$\text{transport of momentum} = -\eta\partial v_x/\partial z.$$

When the temperature T is a function of z, the thermal conductivity κ is defined by

$$\text{transport of thermal energy} = -\kappa\partial T/\partial z.$$

On the other hand, for gases of the same kind at the same temperature and density, the tranport of momentum is proportional to the product between the gradient of average momentum mv_x and l; the transport of thermal energy is proportional to the product between the gradient of average thermal energy c_vT and l' (the constant of proportionality is the same). Hence we have

$$\kappa/\eta = c_v l'/ml.$$

Molecules with large thermal energy move fast on the average and therefore have long free paths. It may thus be understood that f > 1. When the internal degrees of freedom play a role, internal motion of molecules depends little on molecular velocities, so that f for a polyatomic gas is smaller than f = 2.5 for a monatomic gas.

We now split the thermal conductivity κ into κ' due to translational motion and κ'' due to internal motion; the specific heat per molecule is likewise split into c_v'

and c_v''

$$\kappa = \kappa' + \kappa'', \quad c_v = c_v' + c_v''. \tag{10.60}$$

We also assume that the mean free path effective for conduction of internal energy of molecules is equal to that for viscosity:

$$\kappa'' = \eta c_v''/m. \tag{10.61}$$

Combination of (10.60), (10.61), and

$$\kappa' = 5\eta c_v'/2m$$

yields

$$\kappa = \eta(5c_v' + 2c_v'')/2m.$$

Substituting the ratio $\gamma = (c_v + k)/c_v$ between the constant-volume specific heats and $c_v' = 3k/2$, we obtain

$$\kappa = \frac{9\gamma - 5}{4} \eta \frac{c_v}{m}. \tag{10.62}$$

The value of $\mathrm{f} = m\kappa/c_v\eta$ is thus equal to

$$\mathrm{f} = (9\gamma - 5)/4. \tag{10.63}$$

Measured values for many of the nonpolar gases are close to this, known as Eucken's formula.

10.7 $\Omega^{(l, r)}$ for the Lennard-Jones potential

For a specific treatment of transport coefficients in gases, one must calculate the integrals of $\Omega^{(l,r)}$ in (10.27) based on an appropriate intermolecular potential. $\Omega^{(l,r)}$ is a product of the collision cross section and the relative velocity between molecules; it has the meaning of 'resistance against transport.'

Setting the potential between two colliding molecules as $U(r)$, we first derive the expression for the deflection angle (or the scattering angle) θ of the relative velocity when the two molecules collide with the magnitude g of the relative velocity and the impact parameter b.

Let r and θ_1 represent the instantaneous relative position of the approaching molecules in polar coordinates; θ_1 is chosen so that it increases from 0 as shown in Figure 10.4. Denoting the reduced mass of the molecules by m^*, we express the conservation relations of angular momentum and energy as

$$m^* r^2 \frac{d\theta_1}{dt} = m^* g b, \tag{10.64}$$

$$\tfrac{1}{2} m^* \left[\left(\frac{dr}{dt} \right)^2 + r^2 \left(\frac{d\theta_1}{dt} \right)^2 \right] + U(r) = \tfrac{1}{2} m^* g^2. \tag{10.65}$$

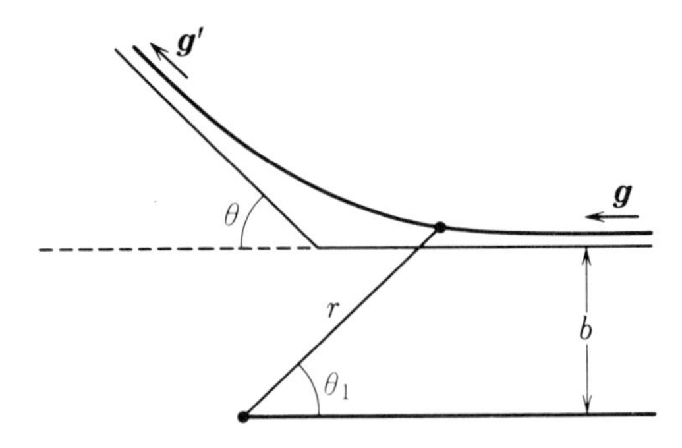

Figure 10.4

Substituting the solution of (10.64) for $d\theta_1/dt$ into (10.65), we obtain

$$\frac{d\theta_1}{dt} = \frac{gb}{r^2}, \quad \frac{dr}{dt} = \left[g^2 - \frac{2U(r)}{m^*} - \frac{g^2 b^2}{r^2} \right]^{1/2},$$

whence

$$\frac{d\theta_1}{dr} = \frac{b}{r^2}\left[1 - \frac{2U(r)}{m^* g^2} - \frac{b^2}{r^2} \right]^{-1/2}.$$

Since the trajectories of the molecules are symmetric with respect to the points of closest approach, the total increment of θ_1 is given by

$$2\int_{r_0}^{\infty} \frac{d\theta_1}{dr}\, dr,$$

where r_0 is the r at the points of closest approach. Since this magnitude corresponds to $\pi - \theta$, the deflection angle θ is

$$\theta = \pi - 2\int_{r_0}^{\infty} \frac{b}{r^2}\left[1 - \frac{2U(r)}{m^* g^2} - \frac{b^2}{r^2} \right]^{-1/2} dr. \tag{10.66}$$

Alternatively, setting $x \equiv b/r$, we have

$$\theta = \pi - 2\int_{0}^{x_0} \left[1 - x^2 - \frac{2}{m^* g^2} U\left(\frac{b}{x}\right) \right]^{-1/2} dx, \tag{10.67}$$

where $x_0 \equiv b/r_0$ is the (smallest) positive root of

$$1 - x^2 - \frac{2}{m^* g^2} U\left(\frac{b}{x}\right) = 0.$$

The collisional cross section, $2\pi b\,db$, between impact parameters b and $b + db$ is equal to the differential cross section for scattering with an angle between θ and $\theta + d\theta$. Thus $d\sigma$ in (10.27) may be replaced by $2\pi b\,db$:

$$\int (1 - \cos^l \theta)\, d\sigma = \int_0^{\infty} (1 - \cos^l \theta)\, 2\pi b\, db.$$

Problem

When both of the colliding molecules are hard spheres, show that

$$\Omega^{(l,r)} = \frac{r_0^2}{4}\left[2 - \frac{1+(-1)^l}{l+1}\right]\left(\frac{\pi kT}{2m^*}\right)^{1/2}(r+1)!$$

where r_0 is the sum of the radii of the two molecules.

Solution

From (10.67) or directly from the point of view of elementary geometry, we have

$$\theta = \pi - 2\sin^{-1}\frac{b}{r_0}, \quad \cos\theta = \frac{2b^2}{r_0^2} - 1 \quad (b \leqslant r_0).$$

When $b > r_0$, of course $\theta = 0$. From this,

$$\int_0^\infty (1 - \cos^l\theta) 2\pi b\,db = \pi r_0^2\left[1 - \frac{1+(-1)^l}{2l+2}\right].$$

Substitution of this in (10.27) yields the required result.

In an 'attracting points-of-mass' model in which the inverse sixth-power attractive potential,

$$U(r) = -\mu r^{-6}, \quad \mu > 0,$$

persists down to an extremely short distance, one has

$$\Omega^{(l,r)} = \tfrac{1}{2}\beta_l\left(\frac{\mu}{2kT}\right)^{1/3}\left(\frac{\pi kT}{2m^*}\right)^{1/2}\Gamma(r + 2 - \tfrac{1}{3}), \quad \beta_1 = 1.9879, \quad \beta_2 = 1.5006.$$

For a repulsive potential of the inverse twelfth power,

$$U(r) = \lambda r^{-12}, \quad \lambda > 0,$$

one likewise has

$$\Omega^{(l,r)} = \tfrac{1}{2}\alpha_l\left(\frac{\lambda}{2kT}\right)^{1/6}\left(\frac{\pi kT}{2m^*}\right)^{1/2}\Gamma(r + 2 - \tfrac{1}{6}), \quad \alpha_1 = 1.1743, \quad \alpha_2 = 0.9466.$$

For the Lennard-Jones potential,

$$U(r) = \frac{\lambda}{r^{12}} - \frac{\mu}{r^6} = U_0\left[\left(\frac{r_0}{r}\right)^{12} - 2\left(\frac{r_0}{r}\right)^6\right] \tag{10.68}$$

one then has a general expression,

$$\Omega^{(l,r)} = 2\left(\frac{\lambda}{2kT}\right)^{1/6}\left(\frac{\pi kT}{2m^*}\right)^{1/2} F_r^l(\zeta), \tag{10.69}$$

$$\zeta \equiv \frac{1}{2kT}\frac{\mu^2}{\lambda} = \frac{2U_0}{kT},$$

Table 10.2 Values of various functions in equation (10.69)

ζ	$F_1^1(\zeta)$	$F_2^1(\zeta)$	$F_2^2(\zeta)$	$F_3^2(\zeta)$
0.000	0.5063	1.4344	1.1564	4.443
0.025	0.4950	1.4073	1.1201	4.310
0.050	0.4912	1.3976	1.1070	4.264
0.075	0.4889	1.3909	1.0982	4.232
0.100	0.4873	1.3859	1.0916	4.207
0.150	0.4857	1.3789	1.0828	4.171
0.200	0.4853	1.3745	1.0779	4.146
0.250	0.4858	1.3720	1.0755	4.129
0.375	0.4897	1.3710	1.0779	4.111
0.500	0.4964	1.3756	1.0883	4.117
0.625	0.5050	1.3842	1.1046	4.141
0.750	0.5150	1.3960	1.1251	4.179
0.875	0.5259	1.4104	1.1486	4.228
1.000	0.5376	1.4270	1.1749	4.286
1.125	0.5499	1.4456	1.2028	4.353
1.250	0.5626	1.4657	1.2321	4.425
1.375	0.5756	1.4872	1.2622	4.503
1.500	0.5887	1.5099	1.2930	4.586
1.625	0.6019	1.5336	1.3240	4.672
1.750	0.6151	1.5581	1.3549	4.761
1.875	0.6283	1.5833	1.3857	4.852
2.000	0.6413	1.6092	1.4162	4.944

which agrees with the case of $-\mu r^{-6}$ in the limit of low temperatures and with the case of λr^{-12} in the limit of high temperatures. The numerical values of the functions $F_r^l(\zeta)$ were computed by Kihara and Kotani (T. Kihara and M. Kotani (1943), *Proc. Phys. Math. Soc., Japan,* **25**, 602) and by Hirschfelder *et al.* (J.O. Hirschfelder, R.B. Bird, and E.L. Spotz (1949), *J. Chem. Phys.*, **17**, 1343), independently; the two results precisely agree with each other. Table 10.2 lists the results of the former computations.

10.8 Interatomic potential for rare gases determined from transport coefficients

The viscosity η of a single-component gas is given from (10.43) and (10.45) by

$$\eta = \frac{5kT}{8\Omega^{(2,2)}}\left[1 + \frac{3}{49}\left(\frac{\Omega^{(2,3)}}{\Omega^{(2,2)}} - \frac{7}{2}\right)^2\right]. \tag{10.70}$$

The thermal conductivity κ of a single-component gas consisting of monatomic molecules is given from (10.54) and (10.55) by

$$\kappa = \frac{3k}{2}\frac{25kT}{16m}\frac{1}{\Omega^{(2,2)}}\left[1 + \frac{2}{21}\left(\frac{\Omega^{(2,3)}}{\Omega^{(2,2)}} - \frac{7}{2}\right)^2\right], \tag{10.71}$$

where m denotes the mass of a molecule.

Choosing the Lennard-Jones potential,

$$U(r) = U_0 \left[\left(\frac{r_0}{r} \right)^{12} - 2 \left(\frac{r_0}{r} \right)^{6} \right],$$

for rare gases, we express $\Omega^{(l,r)}$ as (10.69), i.e.

$$\Omega^{(l,r)} = 2 \left(\frac{U_0}{2kT} \right)^{1/6} r_0^2 \left(\frac{\pi kT}{m} \right)^{1/2} F_r^l(\zeta), \quad \zeta = \frac{2U_0}{kT}.$$

The functions $F_r^l(\zeta)$ have been given in Table 10.2. Substituting these in (10.70), we express the viscosity η in the form

$$\eta = \frac{5}{16} \frac{1}{r_0^2} \left(\frac{mkT}{\pi} \right)^{1/2} \frac{2^{1/3}}{\zeta^{1/6} F(\zeta)}, \qquad \zeta = \frac{2U_0}{kT},$$

$$\frac{1}{F(\zeta)} = \frac{1}{F_2^2(\zeta)} \left[1 + \frac{3}{49} \left(\frac{F_3^2(\zeta)}{F_2^2(\zeta)} - \frac{7}{2} \right)^2 \right];$$

we may similarly treat κ.

When the measured values of η (or κ) are available over a sufficiently wide range of temperatures, those results may be used for determination of the two constants, U_0 and r_0. To do so we may proceed as follows. On transparent graph paper, we plot $-\log\,[\zeta^{1/6}\,F(\zeta)]$ on the ordinate and $-\log \zeta$ on the abscissa. On another sheet of graph paper, measured values are plotted with $\log\,[\eta/(mkT)^{1/2}]$ on the ordinate and $\log T$ on the abscissa. Those two sheets of graphs overlap nicely by parallel translations; the distances between the pairs of coordinate axes determine r_0^2 and U_0/k.

Table 10.3 lists the values of those constants first determined from viscosity by Kihara and Kotani (1943) and the values determined recently from both viscosity and thermal conductivity by Hogervorst.

We note the non-negligible difference between the values in the corresponding pairs, and the similarly non-negligible discrepancy between those values and the values determined from the second virial coefficients (Table 6.1). The principal

Table 10.3 Potential constants determined from transport coefficients

	Kihara and Kotani (1943)		Hogervorst (1971)	
	r_0 (Å)	U_0/k (K)	r_0 (Å)	U_0/k (K)
Ne	3.19	30.4	3.06	43
Ar	3.91	107	3.77	135
Kr	–	–	4.01	193
Xe	4.55	229	4.40	256

T. Kihara and M. Kotani (1943). *Proc. Phys. Math. Soc. Japan*, **25**, 602.
W. Hogervorst (1971). *Physica*, **51**, 77.

reason for these discrepancies may be traced to the limit of accuracy in the Lennard-Jones potential, which leads to differences in the potential constants so determined depending on the temperature ranges of measurements.

For viscosity η, Sutherland's empirical formula has been known for quite some time, that is,

$$\eta = A\frac{\sqrt{T}}{1+S/T}.$$

Here A and S are molecular constants; in particular S is called Sutherland's constant. It has become clear through investigations that our expression for η may be well approximated by

$$\eta \doteqdot \frac{5}{16}\frac{1}{r_0^2}\left(\frac{mkT}{\pi}\right)^{1/2}\frac{1.46}{1+1.10U_0/kT}$$

over the domain of $0.1 \lesssim U_0/kT \lesssim 1$. Thus we may relate the two constants in Sutherland's formula in terms of the two constants in the Lennard-Jones potential.

Chapter 11

Diffusion and Thermal Diffusion in Gases

11.1 Two-component gases

In the previous chapter we treated the transport coefficients of single-component gases, starting with the Boltzmann equation. For a mixed gas, such a mathematical development is quite laborious and the results obtained are generally complicated. Thus in this chapter, paying attention to those specific cases in which simple results are obtainable, we derive semi-intuitively the expressions for the transport coefficients.

For a two-component mixed gas, we distinguish between the components by the suffixes A and B, so that the velocity distribution functions of the molecules are written as $f_A(\mathbf{c}_A, \mathbf{r}, t)$ and $f_B(\mathbf{c}_B, \mathbf{r}, t)$. The number n_A of molecules A in a unit volume is given by

$$n_A(\mathbf{r}, t) = \int f_A(\mathbf{c}_A, \mathbf{r}, t) d\mathbf{c}_A\,;$$

the flow velocity $\mathbf{v}_A$ of molecules A is

$$\mathbf{v}_A = \frac{1}{n_A} \int \mathbf{c}_A f_A d\mathbf{c}_A\,.$$

Similar expressions apply for molecules B.

In the absence of external forces, the velocity distribution functions satisfy the fundamental equations,

$$\frac{\partial f_A}{\partial t} + \mathbf{c}_A \cdot \frac{\partial f_A}{\partial \mathbf{r}} = \left(\frac{\partial_e f_A}{\partial t}\right)_A + \left(\frac{\partial_e f_A}{\partial t}\right)_B, \tag{11.1}$$

$$\frac{\partial f_B}{\partial t} + \mathbf{c}_B \cdot \frac{\partial f_B}{\partial \mathbf{r}} = \left(\frac{\partial_e f_B}{\partial t}\right)_A + \left(\frac{\partial_e f_B}{\partial t}\right)_B, \tag{11.2}$$

which are generalizations of (10.3). Here $(\partial_e f_A/\partial t)_A$ represents the increment of f_A arising from encounters between molecules A; $(\partial_e f_A/\partial t)_B$ represents the increment of f_A arising from encounters between molecules A and molecules B; and similar expressions apply for f_B.

For a mixed gas in thermodynamic equilibrium, $\mathbf{v}_A$ and $\mathbf{v}_B$ are equal and constant; in the coordinate system co-moving with those velocities, f_A and f_B assume the Maxwellian distribution,

$$f_A^{(0)}(\mathbf{c}_A) = n_A \left(\frac{m_A}{2\pi kT} \right)^{3/2} \exp\left(-\frac{m_A c_A{}^2}{2kT} \right), \tag{11.3}$$

$$f_B^{(0)}(\mathbf{c}_B) = n_B \left(\frac{m_B}{2\pi kT} \right)^{3/2} \exp\left(-\frac{m_B c_B{}^2}{2kT} \right), \tag{11.4}$$

where m_A and m_B are the masses of the particles of the two species.

When n_A and n_B vary in space, diffusion takes place. We first consider the case in which the temperature T is uniform. From the condition of uniform pressure we have

$$n_A(\mathbf{r}) + n_B(\mathbf{r}) \equiv n = \text{uniform}. \tag{11.5}$$

The *diffusion coefficient* D in these circumstances is defined through

$$\mathbf{v}_A - \mathbf{v}_B = -D \left(\frac{1}{n_A} \frac{\partial n_A}{\partial \mathbf{r}} - \frac{1}{n_B} \frac{\partial n_B}{\partial \mathbf{r}} \right). \tag{11.6}$$

On account of (11.5), equation 11.6 is equivalent to

$$\mathbf{v}_A - \mathbf{v}_B = -D \frac{n^2}{n_A n_B} \frac{\partial (n_A/n)}{\partial \mathbf{r}}. \tag{11.7}$$

When the temperature T is not uniform, (11.7) is extended as

$$\mathbf{v}_A - \mathbf{v}_B = -\frac{n^2}{n_A n_B} \left[D \frac{\partial (n_A/n)}{\partial \mathbf{r}} + D_T \frac{\partial \ln T}{\partial \mathbf{r}} \right]. \tag{11.8}$$

D_T is called the *coefficient of thermal diffusion*. With the aid of the dimensionless quantity,

$$k_T \equiv D_T / D,$$

called the *thermal diffusion ratio*, the flow of the component A relative to the component B is written as

$$\mathbf{v}_A - \mathbf{v}_B = -\frac{n^2}{n_A n_B} D \left[\frac{\partial (n_A/n)}{\partial \mathbf{r}} + k_T \frac{\partial \ln T}{\partial \mathbf{r}} \right]. \tag{11.9}$$

For the calculation of the coefficient of thermal diffusion, one notes the ratio between $\mathbf{v}_A - \mathbf{v}_B$ and $\partial \ln T / \partial \mathbf{r}$ under the condition of uniform density.

For ordinary diffusion, the flow vector is directed from a place of large concentration to that of small concentration; D is thus always positive. The thermal diffusion ratio, however, can take on either a positive or negative sign. To obtain qualitative knowledge on this sign, let us consider the case that heavy molecules A with sufficiently small concentration are mixed in light molecules B.

The velocity distribution of the light molecules B deviates from the Maxwellian owing to the temperature gradient. With the coordinate axis $+z$ chosen in the direction of increasing temperature, the distribution becomes asymmetric with respect to plus and minus sides of z as shown in Figure 10.3. When a heavy molecule A is stationary at first, fast B molecules collide mainly from the high-temperature domain (right side in the figure), and slow ones mainly from the low-temperature domain. In particular, when both kinds of molecules resemble hard spheres, collisional effects of fast B molecules are large and thus the A molecule begins to drift under the action of the force toward the low temperature domain. This corresponds to $k_T > 0$. Conversely when the intermolecular potential slowly approaches zero as the intermolecular distance increases, collisional effects of slow B molecules are effective; the A molecule drifts toward the high temperature domain. This corresponds to $k_T < 0$. For collisions between neutral molecules the cases with $k_T > 0$ prevail at ordinary temperatures.

11.2 The coefficient of diffusion

We first calculate the expression for the coefficient of diffusion D defined by (11.7) for a two-component gas with uniform temperature, i.e.

$$\mathbf{v}_A - \mathbf{v}_B = -D \frac{n}{n_A n_B} \frac{\partial n_A}{\partial \mathbf{r}} . \tag{11.10}$$

We substitute on the left-hand side of (11.1) the Maxwellian (11.3) in which n_A is replaced by $n_A(\mathbf{r})$; the left-hand side then becomes $\mathbf{c}_A \cdot (\partial \ln n_A/\partial \mathbf{r}) f_A$. Multiplying this by $m_A \mathbf{c}_A$ and carrying out integration with respect to $\mathbf{c}_A$, we obtain $kT\partial n_A/\partial \mathbf{r}$. This quantity must be equal to the integration of the increment of the momentum $m_A \mathbf{c}_A$ of molecules A per unit time and per unit volume due to intermolecular collisions. Collisions between like particles in these circumstances do not produce a direct contribution; they only influence indirectly via the forms of f_A and f_B. We thus have

$$kT \frac{\partial n_A}{\partial \mathbf{r}} = \iiint f_A(\mathbf{c}_A) f_B(\mathbf{c}_B)(m_A \mathbf{c}_A' - m_A \mathbf{c}_A) \mid \mathbf{c}_B - \mathbf{c}_A \mid d\sigma d\mathbf{c}_A d\mathbf{c}_B ,$$

where $\mathbf{c}_A$ and $\mathbf{c}_B$ are the velocities before collision, $\mathbf{c}_A'$ represents the velocities after collision, and we have set the differential cross section with respect to collision between A and B as $d\sigma$.

Ignoring delicate deformations, we may take f_A and f_B to be the Maxwellian around the flow velocities $\mathbf{v}_A$ and $\mathbf{v}_B$. That is, $f_A(\mathbf{c}_A)$ is approximated by (11.3) in which $\mathbf{c}_A - \mathbf{v}_A$ is substituted in place of $\mathbf{c}_A$, i.e.

$$f_A(\mathbf{c}_A) = f_A^{(0)}(\mathbf{c}_A) \left(1 + \frac{m_A}{kT} \mathbf{c}_A \cdot \mathbf{v}_A \right); \tag{11.11}$$

$f_B(\mathbf{c}_B)$ may be approximated by a similar formula.

Introducing now the centre-of-mass velocity of the two-molecule system

$\mathbf{G} \equiv (m_A \mathbf{c}_A + m_B \mathbf{c}_B)/(m_A + m_B)$ and the relative velocity $\mathbf{g} \equiv \mathbf{c}_B - \mathbf{c}_A$, we have

$$\mathbf{c}_A = \mathbf{G} - \frac{m_B}{m_A + m_B}\mathbf{g}, \quad \mathbf{c}_B = \mathbf{G} + \frac{m_A}{m_A + m_B}\mathbf{g}. \tag{11.12}$$

Using further the relative velocity after collision $\mathbf{g}'$ and taking account of $d\mathbf{c}_A d\mathbf{c}_B = d\mathbf{G} d\mathbf{g}$, we may carry out integration with respect to $\mathbf{G}$, to obtain

$$kT\frac{\partial n_A}{\partial \mathbf{r}} = -\frac{n_A n_B}{kT}\iint \left(\frac{m^*}{2\pi kT}\right)^{3/2} \exp(-m^* g^2/2kT)$$
$$\times\, m^*\mathbf{g} \,.\, (\mathbf{v}_A - \mathbf{v}_B) m^*(\mathbf{g} - \mathbf{g}') g\, d\sigma\, d\mathbf{g},$$

where $m^* = m_A m_B/(m_A + m_B)$ is the reduced mass. Integration of $(\mathbf{g} - \mathbf{g}')\mathbf{g} \,.\, (\mathbf{v}_A - \mathbf{v}_B)$ with respect to the directions of $\mathbf{g}$ yields 1/3 of the integration of $(\mathbf{g} - \mathbf{g}') \,.\, \mathbf{g}(\mathbf{v}_A - \mathbf{v}_B)$; on account of this, we find that the right-hand side becomes

$$-\tfrac{16}{3} n_A n_B m^* \Omega^{(1,1)} (\mathbf{v}_A - \mathbf{v}_B),$$

where $\Omega^{(1,1)}$ is the effective cross section between A and B defined in (10.27). On comparing this with (11.10), the coefficient of diffusion D is determined as

$$D = \frac{3}{16}\frac{m_A + m_B}{m_A m_B}\frac{kT}{n\Omega^{(1,1)}}\,. \tag{11.13}$$

Appearance of the number of molecules, $n = n_A + n_B$, per unit volume in the denominator indicates that the coefficient of diffusion is inversely proportional to the pressure at a constant temperature. This aspect is different from the cases of viscosity and thermal conductivity.

The expression obtained above does not take account of subtle deviation of the velocity distribution function from (11.11); it thus corresponds to the first approximation in the previous chapter.

When heavy molecules diffuse in a light gas, it is often the case that concentration of the heavy molecules is thin. In this limit, (11.13) represents an accurate expression. For, the velocities of molecules in the light gas are accurately expressed by the distribution in the form of (11.11) and the velocities of heavy molecules are small. The result in these cases can also be derived in connection with the Brownian motion of heavy molecules.

When the coefficient of diffusion is measured over a sufficiently wide range of temperatures, the potential between molecules A and B can be determined from (11.13) based on those measured values. For example, adopting the potential of the Lennard-Jones type,

$$U_{AB}(r) = U_{0\,AB}\left[\left(\frac{r_{0\,AB}}{r}\right)^{12} - 2\left(\frac{r_{0\,AB}}{r}\right)^{6}\right],$$

we may determine the two constants $U_{0\,AB}$ and $r_{0\,AB}$ with the aid of the values of the function $\Omega^{(1,1)}$ obtained from Table 10.2.

Table 11.1 Potential constants between molecules of different kinds determined from transport coefficients

	From the diffusion coefficient*		Computed values from the potential constants between like particles	
	r_{0AB} (Å)	U_{0AB}/k (K)	r_{0AB} (Å)	U_{0AB}/k (K)
Ne–Ar	3.47	64.5	3.47	67
Ne–Kr	3.62	71.5	3.63	73
Ne–Xe	3.88	73	3.90	73
Ar–Kr	3.94	148	3.90	160
Ar–Xe	4.10	178	4.11	175

*W. Hogervorst (1971). *Physica*, **51**, 59.

Hogervorst (1971) measured the coefficient of diffusion of heavy rare-gas atoms in light rare gases over the temperatures from 300 K to 1400 K; based on those results the values in Table 11.1 were obtained. In addition, Hogervorst determined the constants in the potential between similar rare-gas atoms from viscosity; the values computed from those potential constants with the aid of the connecting formula in Section 6.5 are also listed there for comparison.

The coefficient of diffusion between isotopes is sometimes called the *coefficient of self-diffusion.* Usually for such a measurement a radioactive isotope is chosen as one of the constituents. Neglecting the mass difference between the isotopes and setting the mass of a molecule as m, we may simplify (11.13) as

$$D = \frac{3kT}{8mn\Omega^{(1,1)}} \cdot \tag{11.14}$$

A more accurate expression (corresponding to the second approximation in the previous chapter) may be obtained by multiplying this by

$$1 + \frac{1}{10 + 2\Omega^{(2,2)}/\Omega^{(1,1)}} \left(\frac{\Omega^{(1,2)}}{\Omega^{(1,1)}} - \frac{5}{2} \right)^2 \tag{11.15}$$

11.3 Thermal diffusion between isotopes

The phenomenon of thermal diffusion may be used for separation of isotopes (isotopic elements). The isotopes are different only in their masses, their intermolecular forces being the same; theoretical treatments are thus simplified.

General expressions for the transport coefficients in two-component gases involve the effective cross sections $\Omega_A^{(l,r)}$, $\Omega^{(l,r)}$, and $\Omega_B^{(l,r)}$ between AA, between AB, and between BB; since the intermolecular forces are not different in a mixed gas of the isotopes, we have

$$m_A^{1/2}\Omega_A^{(l,r)} = (2m^*)^{1/2}\Omega^{(l,r)} = m_B^{1/2}\Omega_B^{(l,r)}, \tag{11.16}$$

where m_A and m_B are the molecular masses and $m^* = m_A m_B/(m_A + m_B)$ is the reduced mass. Throughout this section we choose $m_A > m_B$.

Except for hydrogen, relative difference in the masses $(m_A - m_B)/(m_A + m_B)$ between the isotopes is small; the thermal diffusion ratio k_T is proportional to this quantity:

$$k_T \equiv \frac{n_A n_B}{(n_A + n_B)^2} \frac{m_A - m_B}{m_A + m_B} k_T^* . \tag{11.17}$$

The quantity k_T^* is called the *reduced thermal diffusion ratio*; it may be calculated from (11.9) with uniform density, i.e.

$$\mathbf{v}_A - \mathbf{v}_B = -\frac{m_A - m_B}{m_A + m_B} D k_T^* \frac{\partial \ln T}{\partial \mathbf{r}} , \tag{11.18}$$

where D is the coefficient of self-diffusion given by (11.14).

When all the molecules in the gas are the component A, $f_A(\mathbf{c}_A)$ is given by (10.56), that is,

$$f_A^{(0)}(\mathbf{c}_A)\left[1 + a\left(\frac{5}{2} - \frac{m_A c_A^2}{2kT}\right)\left(\frac{m_A}{2kT}\right)^{1/2} \mathbf{c}_A \cdot \frac{\partial \ln T}{\partial \mathbf{r}}\right].$$

Here $f_A^{(0)}(\mathbf{c}_A)$ is the Maxwellian; the coefficient a is

$$a = \frac{15}{16n}\left(\frac{2kT}{m_A}\right)^{1/2} \frac{1}{\Omega_A^{(2,2)}} ,$$

or on account of (11.16) we may also write

$$a = \frac{15}{16n}\left(\frac{kT}{m^*}\right)^{1/2} \frac{1}{\Omega^{(2,2)}} , \tag{11.19}$$

Similarly when all the molecules are B, $f_B(\mathbf{c}_B)$ is given by the same form with the same a:

$$f_B^{(0)}(\mathbf{c}_B)\left[1 + a\left(\frac{5}{2} - \frac{m_B c_B^2}{2kT}\right)\left(\frac{m_B}{2kT}\right)^{1/2} \mathbf{c}_B \cdot \frac{\partial \ln T}{\partial \mathbf{r}}\right].$$

In a homogeneous mixed gas of isotopes A and B, the component A flows with a velocity $\mathbf{v}_A$, and the component B with $\mathbf{v}_B$. Hence $f_A^{(0)}(\mathbf{c}_A)$ must be replaced by $f_A^{(0)}(\mathbf{c}_A - \mathbf{v}_A)$. The velocity distribution then takes the following form:

$$f_A(\mathbf{c}_A) = f_A^{(0)}(\mathbf{c}_A)\left[1 + \frac{m_A}{kT}\mathbf{c}_A \cdot \mathbf{v}_A + a\left(\frac{5}{2} - \frac{m_A c_A^2}{2kT}\right)\left(\frac{m_A}{2kT}\right)^{1/2} \mathbf{c}_A \cdot \frac{\partial \ln T}{\partial \mathbf{r}}\right].$$

This is our first approximation; a similar expression applies for $f_B(\mathbf{c}_B)$.

Now, the relative flow velocity $\mathbf{v}_A - \mathbf{v}_B$ should be determined from the condition that transfer of momentum between the two components vanish on the average. This condition is

$$\iiint f_A(\mathbf{c}_A) f_B(\mathbf{c}_B)(m_B\mathbf{c}_B - m_B\mathbf{c}'_B) \mid \mathbf{c}_B - \mathbf{c}_A \mid d\sigma d\mathbf{c}_A d\mathbf{c}_B = 0,$$

where $\mathbf{c}_A$ and $\mathbf{c}_B$ are the velocities before collision; $\mathbf{c}'_A$ and $\mathbf{c}'_B$ are those after collision. Introducing the centre-of-mass velocity $\mathbf{G}$ and the relative velocities, $\mathbf{g} \equiv \mathbf{c}_B - \mathbf{c}_A$ and $\mathbf{g}' \equiv \mathbf{c}'_B - \mathbf{c}'_A$, we may rewrite the foregoing condition as

$$\iiint f_A(\mathbf{c}_A) f_B(\mathbf{c}_B)(\mathbf{g} - \mathbf{g}') g d\sigma d\mathbf{g} d\mathbf{G} = 0.$$

Performing the integrations we obtain the following result:

$$8\left(\frac{m^*}{kT}\right)^{1/2} (\mathbf{v}_A - \mathbf{v}_B)\Omega^{(1,1)} = -3\frac{m_A - m_B}{m_A + m_B} a(2\Omega^{(1,2)} - 5\Omega^{(1,1)})\frac{\partial \ln T}{\partial \mathbf{r}} .$$

Substituting the expression (11.19) for a in this result and comparing it with (11.18) we obtain

$$k_T^* = \frac{15}{8} \frac{2\Omega^{(1,2)} - 5\Omega^{(1,1)}}{\Omega^{(2,2)}} . \qquad (11.20)$$

Recalling now equation (10.29) for $\Omega^{(l,r)}$, i.e.

$$T\frac{d\Omega^{(1,1)}}{dT} = \Omega^{(1,2)} - \frac{5}{2}\Omega^{(1,1)} ,$$

we find that the thermal diffusion ratio vanishes when $\Omega^{(1,1)}$ is independent of the temperature. That is, for Maxwell's model in which the potential between A and B is inversely proportional to the fourth power of the distance r, no thermal diffusion appears. The phenomenon of thermal diffusion was thus overlooked by Maxwell; it was first clarified theoretically by S. Chapman and D. Enskog (1917). The method later used extensively for enrichment of uranium 235 is one which employs thermal diffusion between uranium hexafluorides, $U^{235}F_6$ and $U^{238}F_6$. The expression (11.20) derived semi-intuitively above is frequently called Kihara's first approximation; despite its simplicity, it has a good accuracy with errors less than 1%.

Problem

When the molecules are hard spheres, compute the value of k_T^*.

Solution

From the problem in Section 10.7, the ratio between the magnitudes of $\Omega^{(1,1)}$, $\Omega^{(1,2)}$, $\Omega^{(2,2)}$ is 1:3:2. From this,

$$k_T^* = 15/16.$$

Incidentally, the magnitude in the second approximation is 1.006 times this value.

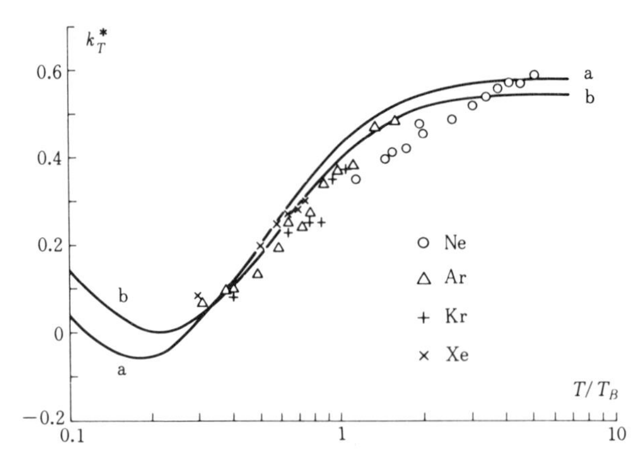

Figure 11.1 Reduced thermal diffision ratio versus the temperature in units of the Boyle temperature. From T. Kihara (1970). Intermolecular Forces. *Physical Chemistry*, Vol. 5

As may be clear from the expression (11.20), the reduced thermal diffusion ratio k_T^* depends sensitively on the functional form of the intermolecular potential. The curves, a and b, in Figure 11.1 are obtained on the basis of

$$U(r) = U_0 \left[\left(\frac{r_0}{r} \right)^{12} - 2 \left(\frac{r_0}{r} \right)^6 \right]$$

and

$$U(r) = U_0 \left[\frac{2}{3} \exp \left\{ 15 \left(1 - \frac{r}{r_0} \right) \right\} - \frac{5}{3} \left(\frac{r_0}{r} \right)^6 \right] ,$$

respectively. For the latter, Mason's table (E.A. Mason (1954), *J. Chem. Phys.*, **22**, 169) has been used. Neither curve is sufficient to explain the measured values completely.

11.4 The rigid convex-body model of polyatomic molecules

When the molecules are not spherically symmetric, the dynamics of collision become very complicated. One of the reasons is involvement of the moments of inertia of the molecules arising from the change in their rotational states before and after collision. Diffusion of heavy molecules in a light gas is an exceptional case, however; dynamical states of heavy molecules moving slowly do not change appreciably in the event of collision with a light molecule. In this section, we briefly discuss such a situation.

As was stated in Section 11.2, when heavy molecules with small molar concentration diffuse in a light gas, (11.13) not only represents the first approximation but also the exact expression. Let the component A be the heavy molecules, and the component B the light molecules. When the molecules A and B are hard spheres with radii a and b and when the attractive force between the molecules A and B is negligible, the expression

$$\Omega^{(1,1)} = \left(\frac{\pi kT}{2m^*}\right)^{1/2} (a+b)^2$$

obtained in the problem of Section 10.7 may be substituted into (11.14), so that the diffusion coefficient D is

$$D = \frac{3}{4}\frac{1}{n}\left(\frac{2\pi kT}{m^*}\right)^{1/2} \frac{1}{4\pi(a+b)^2} \,. \tag{11.21}$$

We can now generalize the model of a heavy molecule from a sphere to a convex body (Kihara 1957). $4\pi(a+b)^2$ is the surface area of a sphere with radius $a+b$. To extend a heavy sphere to a heavy convex body, we may use the surface area of the parallel body of the convex body with thickness b in place of $4\pi(a+b)^2$. This may be understood in the following way. When a light sphere collides with a heavy convex body, the centre of the sphere is reflected at the surface of a parallel body of the convex body with thickness b (Figure 11.2). A given area of surface element on a parallel body reflects on the average a fixed number of spheres, that is, the effect of a surface element of the parallel body is independent of the shape of the parallel body.

Application to the actual molecules may be carried out in the following way. Supposing a core inside a heavy molecule, we let the parallel body of the core with thickness a represent the model of the heavy molecule. The surface area of the parallel body with further thickness b is then, according to Steiner's formula (Section 7.2),

$$S + 2M(a+b) + 4\pi(a+b)^2 \,,$$

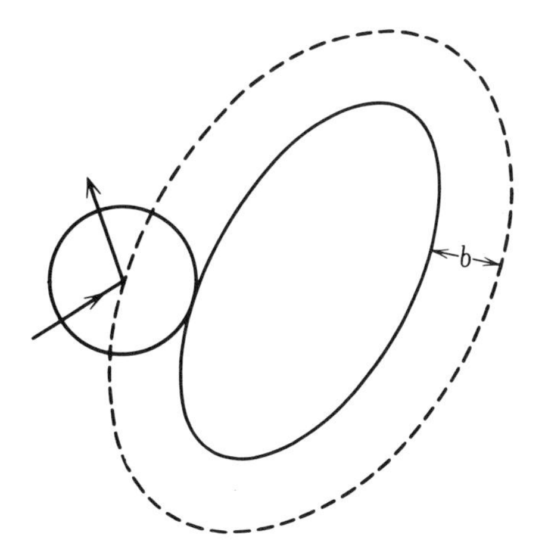

Figure 11.2 Collision between a heavy convex body and a light sphere

where S and M are the surface area and the one-dimensional measure of the core. Consequently, the coefficient of diffusion is expressed by the following formula:

$$D = \frac{3}{4}\frac{1}{n}\left(\frac{2\pi kT}{m^*}\right)^{1/2} [S + 2M(a + b) + 4\pi(a + b)^2]^{-1}. \tag{11.22}$$

The fundamental measures, S and M, of the heavy molecule may be chosen according to or similarly to Table 7.3.

Let us fit (11.22) with the values of the diffusion coefficient for various molecules at 0 °C in hydrogen gas and thereby determine the model constant a. First, from the measured values of the self-diffusion coefficient the radius of an H_2 molecule may be approximated as $b = 1.27$ Å. The values of a are then determined as follows. In units of Å,

N_2 1.52, O_2 1.43, CO_2 1.52, CS_2 2.06, C_6H_6 2.05.

Comparing those values with $\rho_0/2$ in Table 7.3, i.e.

N_2 1.80, O_2 1.60, CO_2 1.65, CS_2 1.80, C_6H_6 1.70,

we find that those molecules which are expected to have large intermolecular attractive forces with H_2 have large values of the ratio $2a/\rho_0$. This is because the effects of increase in the collisional cross section arising from the attractive force are taken into the model constant, a.

11.5 The core potential applied to transport phenomena

As a preparatory consideration to the complex problem of collision between polyatomic molecules, let us determine upper and lower limits of transport coefficients, with the aid of the core potential determined in Chapter 7 and by paying attention to relative magnitudes of the moment of inertia. Specifically we treat viscosity and the coefficient of self-diffusion.

By supposing a convex-body core inside a molecule, the core potential idealizes the intermolecular potential as a function,

$$U(\rho) = U_0 \left[\left(\frac{\rho_0}{\rho}\right)^{12} - 2\left(\frac{\rho_0}{\rho}\right)^{6}\right],$$

of only the shortest distance ρ between the cores. Cores of molecules and the constants such as U_0 and ρ_0 have been summarized in Table 7.3.

If the cross section of collision is calculated using, in place of the convex-body core, a sphere inscribed to the core surface with its centre located at the centre of mass, and by keeping U_0 and ρ_0 unchanged, the result would yield a value smaller than the true cross section. This replacement corresponds to setting the moment of inertia around the centre of mass of the molecule equal to zero.

For a linear molecule such as N_2 and CO_2, the core is a line segment; the inscribed sphere thus becomes a point. Consequently the foregoing replacement means replacing the core potential by the Lennard-Jones potential without changing U_0 and ρ_0.

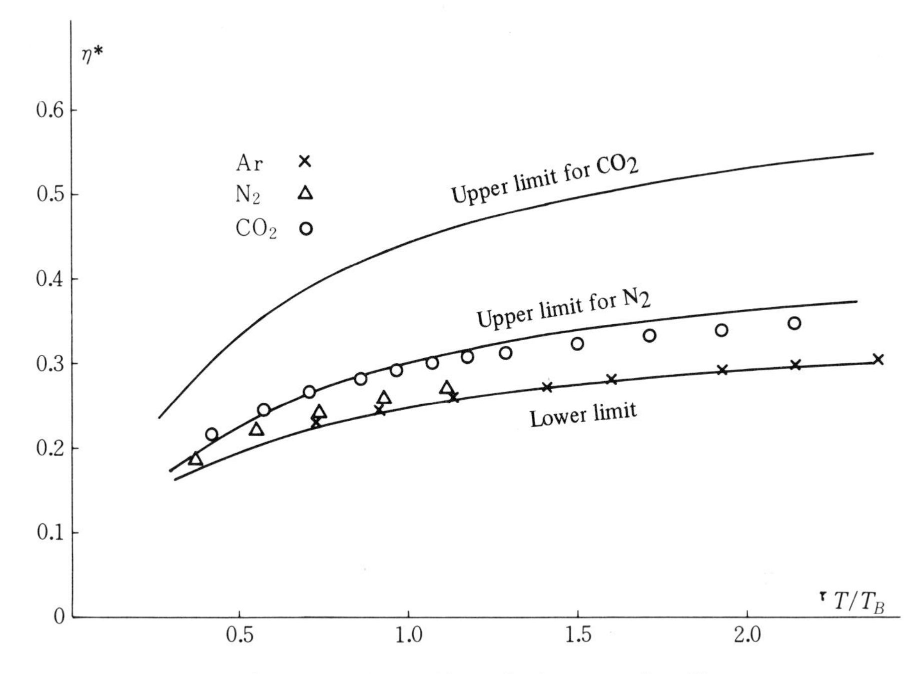

Figure 11.3 Viscosity reduced to a dimensionless quantity. The curves represent the upper and lower limits in the core potential

On the other hand, the upper limit of the collisional cross section may be given by the result of calculation in which the convex-body core is replaced by a spherical core as in Table 7.5. Such a replacement approximately corresponds to assuming that the molecules do not change their rotational states in the event of collision.

For a spherical core, the table of functions necessary for computation for the effective cross sections $\Omega^{(l,r)}$ is fortunately available (J.A. Baker, W. Fock, and F. Smith (1964), *Phys. Fluids,* 7, 897). With the aid of this table, the upper and lower limits of transport coefficients may be obtained by substituting the lower and upper limits of $\Omega^{(l,r)}$, respectively.

Viscosity η of a single-component gas is given by (10.70); the coefficient of self-diffusion, D, by (11.14) and (11.15). Transforming them into dimensionless quantities,

$$\eta^* \equiv b^{2/3}(mkT)^{1/2}\eta, \quad D^* \equiv b^{2/3}nm^{1/2}(kT)^{-1/2}D,$$

we plot them as functions of T/T_B, where T_B is the Boyle temperature and b is the volume defined by (5.2); these have been given in Table 5.1. For N_2 and CO_2, the measured values certainly lie between the upper- and the lower-limit curves, as we see in Figures 11.3 and 11.4. (When the calculated values of the lower limit almost agree, only one curve is shown.) For reference, the calculated curve and measured values for argon are also shown; this curve is based on the potential constants determined from the second virial coefficients.

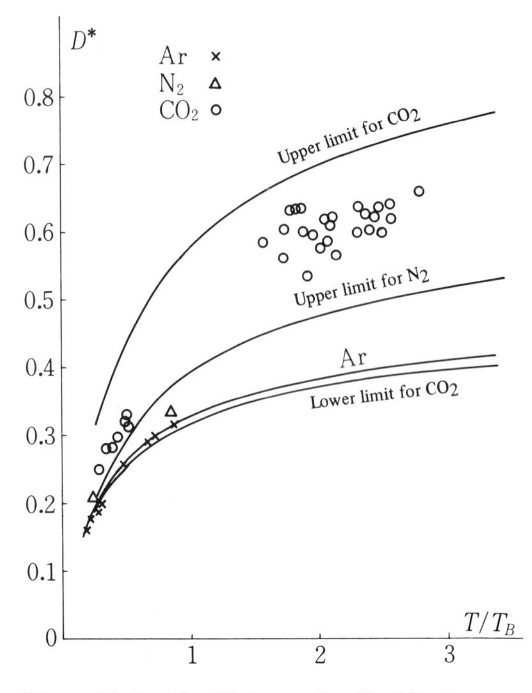

Figure 11.4 Coefficient of self diffusion reduced to a dimensionless quantity

Let us here remark on the following fact. The measured values of viscosity are close to the theoretical curve of the lower limit and many of the measured values of the self-diffusion coefficient lie between the upper- and the lower-limit curves. Replacement of the molecular core by a sphere inscribed in the core represents a relatively good approximation to those collisions in which the trajectories of molecules are substantially deflected; such collisions contribute more heavily to the effective cross section $\Omega^{(1,1)}$ for diffusion than to the effective cross section $\Omega^{(2,2)}$ for viscosity. The fact stated above may be explained in terms of those considerations.

Bibliography

For Chapters 2 and 3 the author is much indebted to:

L. Pauling (1960). *The Nature of the Chemical Bond*, 3rd Ed., Cornell University Press.

Chapters 4–8 are based on the following review articles:

T. Kihara (1953). Virial Coefficients and Models of Molecules in Gases. *Revs. Modern Phys*, **25**, 831.

T. Kihara (1955). Virial Coefficients and Models of Molecules in Gases. *Revs. Mod. Phys.*, **27**, 412.

T. Kihara (1958). Intermolecular Forces and Equation of State of Gases. *Adv. Chem. Phys.*, Vol. 1.

T. Kihara (1963). Convex Molecules in Gaseous and Crystalline States. *Adv. Chem. Phys*, Vol. 5.

A. Dalgarno (1967). New Methods for Calculating Long-Range Intermolecular Forces. *Adv. Chem. Phys.*, Vol. 12.

T. Kihara and A. Koide (1975). Intermolecular Forces and Crystal Structures for D_2, N_2, O_2, F_2, and CO_2. *Adv. Chem. Phys.*, Vol. 33.

The experimental data on virial coefficients are taken from:

J. H. Dymond and E. B. Smith (1969). *Tables of Virial Coefficients of Gases*, Clarendon Press, Oxford.

Sections 5.3 and 7.5 are based on, respectively,

T. Kihara and J. Okutani (1971). Critical Temperature as Accumulation Point of Least Zeroes of Cluster Coefficients. *Chem. Phys. Letters*, **8**, 63.

T. Kihara and K. Miyoshi (1975). Geometry of Three Convex Bodies Applicable to Three-Molecule Clusters in Polyatomic Gases. *J. Stat. Phys.*, **13**, 337.

Chapter 9 is a review of the series of papers:

T. Kihara (1963). Self-Crystallizing Molecular Models. *Acta Cryst.*, **16**, 1119.

T. Kihara (1966). Self-Crystallizing Molecular Models. *Acta Crst.*, **21**, 877.

T. Kihara (1970). Self-Crystallizing Molecular Models. *Acta Cryst.*, **A26**, 315.

T. Kihara (1975). Self-Crystallizing Molecular Models. *Acta Cryst.*, **A31**, 718.

The crystal data are taken from:

W. G. Wyckoff (1971). *Crystal Structures*, 2nd ed. Wiley-Interscience, New York.

The kinetic theory of gases explained in Chapter 10 is a digest of:

S. Chapman and T. G. Cowling (1970). *The Mathematical Theory of Non-Uniform Gases*, 3rd ed. Cambridge University Press.

The following book is also recommended:

J. H. Ferziger and H. G. Kaper (1970). *Mathematical Theory of Transport Processes in Gases*, North-Holland, Amsterdam.

Section 11.3 is taken from

T. Kihara (1975). The Chapman–Enskog and Kihara Approximations for Isotopic Thermal Diffusion in Gases. *J. Stat. Phys.*, **13**, 137.

As regards the molecular beam scattering, which is not treated in the present book, the following articles are useful:

Molecular Beam Scattering, *Faraday Discussions of the Chemical Society, No. 55* (1973).

J. P. Toennies (1974). Molecular Beam Scattering Experiments on Elastic, Inelastic, and Reactive Collisions, *Physical Chemistry*, Vol. 6A, (Academic Press, New York.

This book is concerned with the effects of intermolecular forces mostly in gases and molecular crystals; problems associated with liquids are not treated extensively. Expositions on such topics as computer simulation studies and perturbation theories, which are powerful tools in dealing with liquid states, are somewhat outside the scope of the present volume as set forth in the Preface. Readers are referred to the following monographs for excellent presentation on those subjects: Francis H. Ree (1971), *Computer Calculations for Model Systems*; D. Henderson and J. A. Barker (1971), Perturbation Theories, both in *Physical Chemistry*, vol 8A *Liquid State, Academic Press, New York.*

Bibliographical Notes

Chapters 4, 5, and 6

The famous Lennard-Jones work on the integration of the second virial coefficient was first reported in 1924[1] for the intermolecular potential of the type $\lambda r^{-8} - \mu r^{-4}$; this model was improved to the usual 12–6 potential in 1931.[2]

The three-molecule cluster integral for potentials with inclusion of both repulsive and attractive parts was first evaluated for the square-well potential in 1943;[3] for the Lennard-Jones potential, a series expansion similar to the Lennard-Jones series for the second virial coefficient was obtained in 1948.[4]

The non-additive correction to the potential of the dispersion force was first investigated by Muto[5] and by Axilrod and Teller,[6] each for a simplified atomic model. The general expressions (4.26) and (4.29) are due to Midzuno and Kihara.[7]

The three-molecule cluster integral with nonadditive correction was first investigated by Koba, Kaneko, and Kihara.[8]

The quantum correction to the second virial coefficient in the form of a series expansion in powers of $\hbar^2/mkT$ was first given by Wigner[9] and then by Uhlenbeck and Gropper.[10] Kirkwood[11] supplemented their method by furnishing a more convenient means of obtaining the expansion, which was used by Uhlenbeck and Beth[12] and by Gropper[13] in their recalculations of the quantum correction up to the term proportional to $(\hbar^2/mkT)^2$. The term proportional to the third power together with the correction to the three-molecule cluster integral was calculated by Kihara, Midzuno, and Shizume[14] by use of Husimi's elegant method.[15]

Chapter 7

Spherical-core potentials of intermolecular forces were first proposed in 1947.[16] However, since spherically symmetric potentials are inadequate for such prolonged molecules as carbon dioxide or such flat molecules as benzene, it was desired to generalize the spherical cores to nonspherical rigid bodies.

In 1950 Isihara[17] proved a geometrical identity for systems of two convex bodies and pointed out the utility of the identity for treatment of molecular interactions in uniform gases and liquid solutions. In the same year Hadwiger,[18] a mathematician in Switzerland, gave independently a set of three identities, the first identity corresponding to Isihara's.

Being helped by Ishihara–Hadwiger's identity the present author[19] succeeded in constructing a more general and realistic intermolecular potential for which the second virial coefficient can be evaluated analytically. This is the potential with convex molecular cores described in Chapter 7.

Chapters 10 and 11

The first edition of *The Mathematical Theory of Non-Uniform Gases* by Chapman and Cowling was published in 1939. On the basis of this standard book, computations of transport coefficients for the Lennard-Jones 12–6 potential were first reported by Kihara and Kotani in 1943,[20] and then independently by Hirschfelder, Bird, and Spotz.[21] It should be noted that Kihara and Kotani's paper did not become known to Western scientists until 1949.[22]

The simple formula (11.20) derived somewhat intuitively for the isotopic thermal diffusion in gases was first obtained from

$$k_T^* = \frac{15(2\Omega^{(1,2)} - 5\Omega^{(1,1)})(5\Omega^{(1,1)} + \Omega^{(2,2)})}{\Omega^{(2,2)}(55\Omega^{(1,1)} - 20\Omega^{(1,2)} + 4\Omega^{(1,3)} + 8\Omega^{(2,2)})}$$

given in Chapman and Cowling's book by disregarding smaller quantities.[23]

References

1. J. E. Lennard-Jones (1924). *Proc. Roy. Soc., London*, A **106**, 463.
2. J. E. Lennard-Jones (1931). *Proc. Phys. Soc., London*, A **43**, 461.
3. T. Kihara (1943). *Nippon Sugaku-Buturigaku-kaisi*, **17**, 629.
4. T. Kihara (1948). *J. Phys. Soc., Japan*, **3**, 265.
5. Y. Muto (1943). *Nippon Sugaku-Buturigaku-kaisi*, **17**, 629.
6. B. M. Axilrod and E. Teller (1943). *J. Chem. Phys.*, **11**, 299. B.M. Axilrod (1951). *J. Chem. Phys.*, **19**, 719.
7. Y. Midzuno and T. Kihara (1956). *J. Phys. Soc., Japan*, **11**, 1045.
8. S. Koba, S. Kaneko, and T. Kihara (1956). *J. Phys. Soc., Japan*, **11**, 1050.
9. E. Wigner (1932). *Phys. Rev.*, **40**, 749.
10. G. E. Uhlenbeck and L. Gropper (1932). *Phys. Rev.*, **41**, 79.
11. J. G. Kirkwood (1933). *Phys. Rev.*, **44**, 31.
12. G. E. Uhlenbeck and E. Beth (1936). *Physica*, **3**, 729.
13. L. Gropper (1936). *Phys. Rev.*, **50**, 963.
    L. Gropper (1937). *Phys. Rev.*, **51**, 1108.
14. T. Kihara, Y. Midzuno, and T. Shizume (1955). *J. Phys. Soc., Japan*, **10**, 249.
15. K. Husimi (1940). *Proc. Phys. Math. Soc., Japan*, **22**, 264.
16. T. Kihara (1947). *Nippon Buturigaku-kaisi*, **2**, 11.
17. A, Isihara (1950). *J. Chem. Phys.*, **18**, 1446.
18. H. Hadwiger (1950). *Mh. Math.*, **54**, 345. See H. Hadwiger (1955). *Altes and Neues über konvexe Körper*, Birkhäuser Verlag, Basel und Stuttgart.
19. T. Kihara (1953). *Revs. Modern Phys.*, **25**, 831.
20. T. Kihara and M. Kotani (1943). *Proc. Phys. Math. Soc., Japan*, **25**, 602.
21. J. O. Hirschfelder, R. B. Bird, and E. L. Spotz (1948). *J. Chem. Phys.*, **16**, 968.
22. J. O. Hirschfelder, R. B. Bird, and E. L. Spotz (1949). *J. Chem. Phys.*, **17**, 1343.
23. T. Kihara (1949). *Imperfect Gases*, Asakura Bookstore, Tokyo. English translation by U.S. Air Force Office of Research, Wright-Patterson Air Force Base. See also Reference 19.

Molecule Index

Note: Numbers in this index refer to *section* numbers.

Subject Index